PHYTOMEDICINES, HERBAL DRUGS, AND POISONS

PHYTOMEDICINES, HERBAL DRUGS, AND POISONS

Ben-Erik van Wyk

Michael Wink

The University of Chicago Press
Chicago and London

Kew Publishing
Royal Botanic Gardens, Kew

Conceptualized and developed by Briza Publications, South Africa www.briza.co.za

Copyright © 2014 in text: Ben-Erik van Wyk and Michael Wink
Copyright © 2014 in photographs: Ben-Erik van Wyk and individual photographers listed
Copyright © 2014 in published edition: Briza Publications

All rights reserved. No part of this publication may be reproduced or transmitted in any form or by any means without written permission of the copyright holders.

> **IMPORTANT WARNING**
> This book is a scientific review of medicinal and poisonous plants and not a medical handbook. Some of the plants described may result in death, serious intoxication, severe allergies and other harmful effects. None of the statements in this book can be interpreted as a recommendation to experiment with any of the plants. Neither the authors nor the publishers can be held responsible for claims arising from the mistaken identity of plants or their inappropriate use. Do not attempt self-diagnosis or self-treatment. Always consult a medical professional or qualified practitioner.

Joint publication with the Royal Botanic Gardens, Kew, Richmond, Surrey TW9 3AB, UK.
www.kew.org
and
The University of Chicago Press, Chicago 60637
The University of Chicago Press, Ltd., London

23 22 21 20 19 18 17 16 15   1 2 3 4 5

ISBN-13: 978-0-226-20491-8 (cloth)
ISBN-13: 978- 978-0-226-20507-6 (e-book)
Royal Botanic Gardens, Kew ISBN-13: 978 1 84246 515 8 (cloth)

The views expressed in this work are those of the individual authors and do not necessarily reflect those of the publisher or of the Board of Trustees of the Royal Botanic Gardens, Kew.

Library of Congress Cataloging-in-Publication Data

Van Wyk, Ben-Erik, author.
  Phytomedicines, herbal drugs, and poisons / Ben-Erik Van Wyk and Michael Wink.
     pages cm
  Includes index.
  ISBN 978-0-226-20491-8 (cloth : alk. paper) — ISBN 978-0-226-20507-6 (e-book) 1. Herbs—Therapeutic use. 2. Medicinal plants. 3. Botany, Medical. I. Wink, Michael, author. II. Title.
  RM666.H33V367 2015
  615.3'21—dc23
                                                                    2014046645

∞ This paper meets the requirements of ANSI/NISO Z39.48-1992 (Permanence of Paper).

DOI: 10.7208/chicago/9780226205076.001.0001

Project manager: Reneé Ferreira
Cover design: Lauren Smith
Inside design and typesetting: Alicia Arntzen, The Purple Turtle Publishing CC
Reproduction: Resolution Colour, Cape Town
Printed and bound by Tien Wah Press (Pte.) Ltd, Singapore

Contents

Preface	6
Introduction	7
Traditional systems of medicine	8
Phytomedicines	16
Functional foods and nutraceuticals	18
Plant-derived chemical compounds	20
Mind-altering drugs and stimulants	22
Plant poisons	24
Plant parts used	28
Dosage forms	30
Methods and routes of administration	34
Extraction and analysis of compounds	36
Quality control and safety	40
Efficacy of herbal medicine	42
Pharmacological and toxic effects	44
Regulation and legal aspects	46
Overview of secondary metabolites	48
Photographic index of plants	92
Glossary	274
Further reading	285
Acknowledgements and photo credits	288
Index	289

Preface

Aide-mémoire to the Medicinal Plant Sciences or *Natural Products in a Nutshell* could have been equally appropriate titles for this bird's eye view of medicinal and poisonous plants.

The aim is to present the reader with a compact, fully illustrated, multilingual and user-friendly reference guide. The book covers 360 species of commercially relevant and well-known medicinal plants, including those used for their poisonous or mind-altering activities. It also briefly explains the basic concepts related to the botany, chemistry, pharmacology and use of these plants.

There are few books that cover the entire spectrum of medicinal, poisonous and mind-altering plants, a wide and complicated field of study. Not only are there thousands of plant species that are used in one way or another, but also many different medicinal systems and cultural groups that each have their own *materia medica* and their own ways of using plants. This makes for a fascinating and never-ending scientific exploration, to discover and to learn.

At the request of readers and publishers, we have attempted to condense a large volume of data and concepts into a limited number of pages, in order to present an affordable yet colourful summary of the most important facts relating to medicinal and poisonous plants from all corners of the earth.

This book should be viewed as a convenient and user-friendly starting point (a desk-top reference guide) to get quick, scientifically accurate answers to basic questions. Those who want to delve deeper into the subject can refer to our two other, more comprehensive reviews: *Medicinal Plants of the World* and *Mind-altering and Poisonous Plants of the World*, as well as the many scientific references that are cited there. In our fast-moving modern world, knowledge has become freely available on an unprecedented scale. There are excellent books on almost any conceivable aspect of medicinal botany and hundreds of thousands of scientific papers describing the details of chemical studies and pharmacological evaluations of plant compounds and extracts. Added to this is the worldwide web, where huge amounts of data can be accessed instantaneously. These sources generally provide the long answers, not the short ones. Our target audience covers the full range of readers: interested lay people who may want to use the book as an illustrated encyclopaedia, students of botany and pharmacology who need to prepare for an examination, professional persons working in the commercial environment and even academics and researchers who want to save time and need a quick reference guide and mnemonic aid.

Ben-Erik van Wyk and Michael Wink
1 September 2014

Introduction

This book is intended to be a handy desktop reference book to well-known medicinal and poisonous plants of the world. It is aimed at providing health care professionals, pharmacists, doctors, students and all other interested persons with quick answers to basic questions about medicinal and poisonous plants.

Fundamental concepts, terms and methods relevant to the subject are briefly defined and discussed in short introductory chapters. The aim is to provide scientifically rigorous and accurate information in a study field that is rapidly changing and adapting to the modern way of life.

The use of medicinal, mind-altering and poisonous plants is often associated with folk medicine, practised in distant rural areas where access to modern health care is not available. However, the popularity of natural remedies and botanicals in the modern world cannot be denied, as is seen in the rapid growth of over-the-counter medicines, dietary supplements and functional foods. Many people are taking control of their health by eating a balanced diet, getting regular exercise and using natural medicines and supplements that may help with the prevention and cure of ailments and imbalances. It therefore seems likely that basic knowledge about plant products and their chemical constituents will become increasingly important in the future, as we strive towards better health and a longer, happier life.

There are still many questions about the safety and efficacy of plant-derived products. Some of the answers can be found in this book. It is, however, likely that modern science will not only make dramatic new discoveries in health care, but also provide a deeper understanding of age-old principles that seemed implausible from a reductionist perspective. It took science more than 200 years to discover why limes (lemons) can prevent scurvy; it is not unreasonable to expect that profound new insights still lie ahead in the distant future, despite our best efforts to apply modern technology and the principles of science. When it comes to biology and human health, we know what we do not yet know (and therefore try to find answers to our questions through scientific research) but we also do not know what we do not yet know (and therefore eagerly await ground-breaking discoveries that will allow us to ask the right questions).

Mallow (*Malva sylvestris*): source of anthocyanin pigments

Traditional systems of medicine

The majority of people on earth still rely on traditional medicine for their primary health care needs. Modern allopathic medicine not only co-exists in parallel to the systems from which it was derived, but is often enriched by new discoveries based on ancient knowledge and experience. In general, traditional herbal remedies are used to alleviate the symptoms of chronic and self-terminating illnesses, while allopathic medicines are called upon in case of serious and acute health conditions.

Ancient origins

There is evidence that primates such as chimpanzees and gorillas sometimes ingest particular plants not as food but for their medicinal value. The use of plants as medicine may therefore have a very long history. Recent evidence from southern Africa shows that human abstract thinking dates back to at least 140 000 years ago. This means that most of the history of how medicines developed was never recorded. It is likely that a lengthy process of trial and error resulted in some remedies being rejected as ineffective and perhaps dangerous, while others became important cures. The results of these experiments were no doubt passed on verbally from one generation to the next. In the absence of written records and to ensure maximum mnemonic value, important elements of the cure (i.e., the diagnosis of the ailment, the identity of the plant and the methods of administration) may have been intricately linked to one another within mythological stories, songs or poems. It is also likely that superstition and magic played important roles as ways in which people without scientific insights tried to make sense of what they observed. For example, disease is often associated with evil spirits, which is quite understandable if you have no access to a microscope. The act of "chasing away evil spirits" is almost certainly equivalent to our modern-day practice of using disinfectants and antiseptics.

Traditional medicine is also dynamic and adaptive, as can be seen by the rapid incorporation of recently introduced plant species into the *materia medica*. The process of trial and error was sometimes guided by the "doctrine of signatures", based on the belief that the Creator has provided the plants themselves with clues as to how they should be used. Milky latex, for example, may indicate therapeutic value in promoting lactation; red sap is associated with blood and may suggest efficacy in treating menstrual ailments; yellow sap suggests value as cholagogue, to increase or decrease bile flow, and so on. In traditional cultures there is not such a sharp distinction between food and medicine. Some products are eaten more for their health benefits than for their nutritional value; others are used not to cure any ailments but to prevent them in the first place. It is very likely that all medicines were originally eaten or chewed, as can still be seen in hunter-gatherer communities. Dosage forms such as infusions, decoctions and tinctures must have been a much later development. Some plants were used for ritual and religious purposes, especially those with hallucinogenic properties that provided insights into other realms and other worlds. Ancient systems also incorporate mental health, harmony and balance as important underlying principles of a good life.

African medicine

African Traditional Medicine probably dates back to the origins of our species and represents the most diverse but also the least systematised and most poorly documented of all medicinal systems. There are many regional differences, reflecting the extreme biological and cultural diversity of sub-Saharan Africa, including local plant endemism and local cultural customs. Common to all is holism, in which both body and mind are treated: the underlying psychological basis of the illness is first attended to, after which herbs and other medicines are prescribed to alleviate the symptoms. The ancient practices of the click-speaking people of southern Africa are particularly interesting, not only because they represent the most ancient of human cultures, but also because their traditional home is an area of exceptional plant endemism. In South Africa, an integration of Khoi-San and Cape Dutch healing methods has resulted in a distinct and unique healing system, for which the name **Cape Herbal Medicine** was recently proposed. The remarkably diverse *materia medica* typically includes general tonics, fever remedies, sedatives, stomachics, diuretics, laxatives and many wound-healing plants. Tropical Africa and especially West and East Africa represent a rich diversity of medicinal plants and human cultures. Examples of locally important medicinal plants in Ethiopia include *Echinops kebericho*, *Embelia schimperi*, *Glinus*

San rock art showing aloes

San healing dance

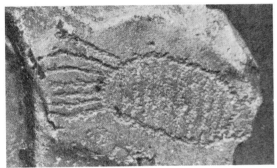

Ancient rock engraving showing bushman poison bulb (*Boophone disticha*)

Juice resembling blood (*Pelargonium antidysentericum*)

Coffee (*Coffea arabica*) – the most popular of all stimulant beverages

Khat (*Catha edulis*) – wrapped in banana leaves to keep fresh

lotoides, *Hagenia abyssinica*, *Lepidium sativum*, *Moringa stenopetala*, *Phytolacca dodecandra*, *Ruta chalepensis* and *Taverniera abyssinica*. The commercially most relevant African medicinal plants have been described in a recent *African Herbal Pharmacopoeia*. It includes *Acacia senegal* (gum arabic), *Agathosma betulina* (buchu), *Aloe ferox* (Cape aloe), *Artemisia afra* (African wormwood), *Aspalathus linearis* (rooibos tea), *Boswellia sacra* (frankincense), *Commiphora myrrha* (myrrh), *Harpagophytum procumbens* (devil's claw), *Hibiscus sabdariffa* (hibiscus or roselle), *Hypoxis hemerocallidea* (African potato), *Ricinus communis* (castor oil plant) and *Prunus africana* (African cherry or red stinkwood). There are many hunting poisons [e.g. *Adenium obesum* (desert rose), *Boophone disticha* (bushman poison bulb)], ordeal poisons [e.g. *Physostigma venenosum* (calabar bean), *Erythrophleum suaveolens* (ordeal tree)] and stimulants [e.g. *Catha edulis* (khat), *Coffea arabica* (coffee) and *Sceletium tortuosum* (kanna or *kougoed*)].

European medicine

European medicine or Galenic medicine has a recorded history dating back to Hippocrates (460–377 BC), Aristotle (384–322 BC) and especially Galen (AD 131–199). The system was based on the four elements (earth, air, fire and water), corresponding with cold, heat, dampness and dryness and also with four humours (blood, phlegm, black bile and yellow bile) and four temperaments (respectively sanguine, phlegmatic, melancholic and choleric). Herbs were used to restore balance but sometimes drastic measures such as bloodletting and purging were used. The system was strongly influenced by what is considered to be the first European herbal, namely *De Materia Medica*, written by the Greek physician Dioscorides in the first century AD. Other famous names include Hildegard of Bingen (1098–1179) and the Swiss alchemist Paracelsus (1493–1541), who is remembered for recognising that the distinction between medicine and poison is only a matter of dose. Amongst the most famous herbals (medicinal handbooks) are the *Historia Stirpium* (1542) and *New Kreüterbuch* (1543) by the German physician Leonhart Fuchs, the *Kruydtboeck* (1581) by the Flemish botanist Matthias de Lobel, the *Herball* (1597) by the English horticulturalist John Gerard and *The English Physician* (1652) by the English pharmacist Nicholas Culpeper.

Herbal medicines are still widely used in countries such as Germany, Austria, France, Italy, Great Britain and Switzerland as natural alternatives to synthetic chemicals, or as supportive treatments. The system incorporates remedies from many parts of the world and is now highly regulated (e.g. through modern pharmacopoeias). Crude drugs are still widely used but sophisticated phytomedicines are becoming increasingly popular because their safety and efficacy have been proven through clinical studies. Commercialised European herbal medicines include *Arnica montana* (arnica), *Drimia maritima* (squill), *Matricaria chamomilla* (chamomile), *Silybum marianum* (milk-thistle), *Urtica dioica* (nettle), *Valeriana officinalis* (valerian) and many others. Famous poisons, formerly used for suicide, murder, execution and political purposes, include *Aconitum napellus* (aconite), *Conium maculatum* (poison hemlock) and *Veratrum album* (white hellebore). The most famous aphrodisiac and hallucinogen is *Atropa belladonna* (deadly nightshade). Pure chemical compounds originally derived from European plants that are used in modern medicine include aspirin and atropine.

Several more holistic approaches to health care had their origins in Europe, including aromatherapy, homoeopathy, anthroposophical medicine and Bach flower remedies. It is interesting and surprising that these somewhat controversial alternative systems had their origins in a region that is dominated by the Western allopathic system with its strong emphasis on evidence-based medicine.

Aromatherapy is a healing system that uses volatile oils as inhalations, massages, baths and perfumes to treat the symptoms of disease and to maintain health. The term aromatherapy was first used in 1928 by the French chemist René Gattefossé. Perhaps the oldest form of aromatherapy was practised by the Sanqua (bushmen) of southern Africa, who habitually massaged themselves with powdered aromatic bushes (*san*) mixed with fat (hence *Sanqua*, -*qua* meaning people). Frankincense and myrrh are still widely used as perfumes and aromatic products, in the same way as was done in ancient Egypt. Essential oils can be absorbed through the skin and mucosa and they are known for their antiseptic, anti-inflammatory and spasmolytic effects. As is the case with perfumes, they are likely to stimulate the mind and mood and along this route also contribute to the restoration of good health and vigour.

Homoeopathy (or homeopathy) is a system of healing proposed by Samuel Hahnemann in Leipzig, Germany, between 1811 and 1820. It is based on the theory that substances can be ingested in very dilute form to treat illnesses associated with the symptoms that would be produced by high doses of the same substances. If a high dose is emetic, for example, then a dilute dose is used as anti-emetic. The word comes from the Greek *homoios* (like) and -*pathy* or *patheia* (suffering). The number of times a mother tincture has been diluted ten-fold (through the so-called process of potentising) is stated after the name of the medicine. For example *Arnika* D3 means a three-times ten-fold dilution; it is considered to be weaker than the more potentised *Arnika* D30. Potentising involves vigorous shaking of the solution with the idea that this action releases immaterial forces from the extracts. Products such as minerals (e.g. arsenic, sulfur) and animal products (e.g. bee venom) are also used.

Anthroposophical medicine is based on principles proposed by the Austrian philosopher and

Examples of medicinal and poisonous plants: benzoin (*Styrax benzoin*), wormseed goosefoot (*Chenopodium ambrosioides*), winter bark (*Canella winterana*), poison sumac (*Rhus vernix*), quassia (*Quassia amara*) and potato (*Solanum tuberosum*)

once again gained in popularity but are mostly viewed (and indeed regulated) as dietary supplements (called "botanicals") rather than medicines in their own right. No direct medical claims are allowed and the products are typically sold as over-the-counter health products, used as supportive treatments. Some, however, have been subjected to clinical studies and are regarded as phytomedicines, used in much the same way as allopathic medicines. Examples of well-known medicinal plants are *Aralia racemosa* (American spikenard), *Capsicum frutescens* (chilli, cayenne), *Cimicifuga racemosa* (black cohosh), *Echinacea purpurea* (echinacea), *Gaultheria procumbens* (wintergreen), *Hamamelis virginiana* (witch hazel), *Hydrastis canadensis* (goldenseal), *Lobelia inflata* (Indian tobacco), *Mahonia aquifolium* (Oregon grape) and *Serenoa repens* (saw palmetto). Poisonous plants include *Gelsemium sempervirens* (yellow jasmine) and the infamous poison ivy (*Rhus toxicodendron*). Damiana (*Turnera diffusa*) is an example of a well-known stimulant.

Phytomedicines

Phytomedicines are plant parts or plant-derived natural substances that are used to treat ailments or alleviate their symptoms (*phyto* is Latin for "plant") in the same way as conventional drugs. The concept of phytomedicine is usually associated with products that have been adequately tested for safety and efficacy so that the consumer can expect real therapeutic benefits without serious risk. In contrast, traditional medicine or folk medicine may be equally effective but there is some degree of uncertainty about its ability to cure the intended ailment or alleviate its symptoms. The uncertainty may arise from questions about the level (concentration) of active ingredients, the proper dose level and its frequency and the most effective method of administration.

In rural societies, roughly 10% of all plants are typically used to some extent for medicinal purposes. This means that 25 000 plant species out of the estimated global total of 250 000 may already have empirical evidence for their therapeutic value. Scientists are continuously exploring the chemistry and pharmacological activities of poorly known medicinal plants in a quest to discover new remedies or to explain the modes of action of these plants. Through this process, the most promising plants (in terms of safety and efficacy) gradually become more popular and may end up as commercialised over-the-counter products.

Of the many traditional medicines that are used worldwide, relatively few have been subjected to rigorous human studies. The term phytomedicine should perhaps be reserved for standardised products (extracts or preparations) that have convincing proof of safety and efficacy in the form of at least one controlled clinical trial with a favourable outcome.

The following is a list of the best-known plant products that meet the criteria for being regarded as phytomedicines. Pure chemical entities are not really phytomedicines but are used in the same way as chemical compounds in prescription medicines (see page 20).

ashwagandha (*Withania somnifera*); roots: ergostane steroidal aglycones (withanolides), alkaloids (physical and psychological stress)
barley (*Hordeum vulgare*); fruit husks (cereal): β-glucans (reducing serum (LDL) cholesterol and controlling diabetes)
black cohosh (*Actaea racemosa*/*Cimicifuga racemosa*); rhizome and roots: triterpenoids (menopausal symptoms)
bloodroot (*Sanguinaria canadensis*); rhizome: sanguinarine (dental plaque, gingivitis)
bugleweed (*Lycopus europaeus*); herb: depsides of hydrocinnamic acids (antithyrotropic)
butterbur (*Petasites hybridus*); rhizomes: sesquiterpenes (petasin, isopetasin) (chronic headache and migraine)
chamomile (*Matricaria chamomilla*); flower heads: essential oil with α-bisabolol (anti-inflammatory, antibiotic)
chaste tree (*Vitex agnus-castus*); standardised extract from fruit: diterpenes (premenstrual stress syndrome, dysmenorrhoea, corpus luteum deficiency, mastalgia)
chilli pepper (*Capsicum* species); fruits: capsaicin (topical pain relief)
Chinese mock-barberry (*Schisandra chinensis*); fruits: lignans (schizandrin A, gomisin) (adaptogenic tonic, increased resistance to stress, hepatoprotective)
Chinese wormwood (*Artemisia annua*); herb: sesquiterpene lactones (artemisinin and derivatives) (antimalarial)
devil's claw (*Harpagophytum procumbens*); secondary roots: iridoid glycosides (low back pain)
echinacea (*Echinacea* species); whole herb and/or root: polysaccharides, alkamides, polyacetylenes, caffeic acids (colds, infections of the respiratory and urinary tracts)
evening primrose (*Oenothera biennis*); seed oil: gamma-linoleic acid (atopic eczema, pruritus, inflammation of skin)
fenugreek (*Trigonella foenum-graecum*); seeds: mucilage, steroidal saponins, alkaloids, peptides (hypoglycaemic, cholesterol-lowering)
feverfew (*Tanacetum parthenium*); herb: sesquiterpene lactones (migraine prophylactic)
flax (*Linum usitatissimum*); seeds: fatty acids, mucilage, lignans (managing high cholesterol and diabetes)
garlic (*Allium sativum*); bulbs: sulfur-containing compounds (antibiotic, lipid-lowering, platelet aggregation inhibiting)
ginger (*Zingiber officinale*); rhizome: pungent gingerols, diterpene lactones, sesquiterpenes (antiemetic, post-operative nausea and travel sickness)
ginkgo (*Ginkgo biloba*); leaf: flavonoids, diterpene lactones (ginkgolides) (cerebrovascular insufficiency and Alzheimer's dementia)

Examples of phytomedicines: evening primrose, garlic, valerian and echinacea

ginseng (*Panax ginseng*); root: triterpenoid saponin (ginsenosides) (enhancement of mood, performance, immune response and convalescence)

griffonia (*Griffonia simplicifolia*); seeds: indole alkaloids (5-hydroxytryptophan) (neurological and psychiatric disorders)

hawthorn (*Crataegus monogyna*); leaves and flowers: procyanidins and flavonoids (cardiac insufficiency; heart rhythm disorders)

horse chestnut (*Aesculus hippocastanum*); seeds: triterpene saponins (venous and lymphatic insufficiency)

Iceland moss (*Cetraria islandica*); whole herb (thallus): polysaccharides (dry cough; inflammation of the mouth and throat)

Indian pennywort (*Centella asiatica*); whole herb: asiaticoside (wound treatment, prevention of scar tissue formation)

ivy (*Hedera helix*); leaf: saponins (cough, bronchitis, chronic catarrh)

kava kava (*Piper methysticum*); rhizome with roots: styrylpyrones (anxiolytic; anxiety, sleep disturbances, stress)

large cranberry (*Vaccinium macrocarpon*); fruit juice/extracts: acids, polyphenols, tannins, procyanidins, anthocyanins (urinary tract infections)

milk-thistle (*Silybum marianum*); fruits: silymarin (mixture of flavanolignans) (cirrhosis, chronic liver inflammation, liver damage)

mistletoe (*Viscum album*); herb: special extract: lectins, viscotoxins, polysaccharides [degenerative inflammation of joints, malignant tumours (palliative treatment)]

onion (*Allium cepa*); bulbs: sulfur-containing compounds (appetite loss, arteriosclerosis)

pale purple coneflower (*Echinacea pallida*); roots: polysaccharides, alkamides, polyacetylenes, caffeic acids (colds and influenza)

peppermint (*Mentha ×piperita*); leaf/essential oil: menthol (digestive ailments, catarrh, irritable bowel syndrome)

purple coneflower – see echinacea

pygeum – see red stinkwood

red stinkwood (*Prunus africana*); bark: phytosterols (benign prostate hyperplasia)

saw palmetto (*Serenoa repens*); fruits: phytosterols (benign prostate hyperplasia)

St John's wort (*Hypericum revolutum*); whole herb: hypericin, hyperforin (mild depression)

stinging nettle (*Urtica dioica*); roots: polysaccharides, a lectin, sitosterols and their glycosides (benign prostate hyperplasia, urological ailments)

tea tree (*Melaleuca alternifolia*); essential oil: terpinen-4-ol, α- and γ-terpinene and 1,8-cineole (acne, fungal infections)

umckaloabo (*Pelargonium sidoides*); root: coumarins (umckalin), tannins, flavonoids (bronchitis, immune stimulant)

valerian (*Valeriana officinalis*); roots: valerenic acid, essential oil, sedative; valtrate, didrovaltrate (minor nervous conditions)

white willow (*Salix alba*); bark: salicin (inflammation, rheumatism, pain)

winter cherry – see ashwagandha

witch hazel (*Hamamelis virginiana*); bark and leaves: tannins (varicose veins, haemorrhoids, diarrhoea)

yellow gentian (*Gentiana lutea*); rhizome and root: bitter secoiridoids (digestive disorders and loss of appetite; bitter tonic)

Functional foods and nutraceuticals

In old and holistic traditional healing systems such as Ayurveda and Chinese Traditional Medicine there has never been a strong distinction between medicine and food: some food items are eaten more for their medicinal than nutritional value. The modern trend in Western countries has been to develop and market functional foods and dietary supplements in order to escape the very strict regulations that apply to medicine. Dietary supplements, for example, do not need approval from the USA Food and Drug Administration but no medical claims may be made. Nutraceuticals can be defined as food items that are eaten for their extra health benefits, in addition to their basic nutritional value. The claims that are typically made include the alleviation of the symptoms of chronic diseases, the prevention of disease, improvement in general health, the delay of ageing, increased life expectancy and the support of the structure and/or function of the body. Functional foods resemble conventional food in appearance and use but often contain added ingredients that enrich the nutritional value and/or provide added medicinal or physiological benefits. In Japan, products sold as functional food have to be used in their natural form (tablets or powders are not allowed), they must be consumed as part of a normal daily diet and they should have some value in preventing or controlling diseases.

Prebiotics are non-digestible fibre compounds (mostly cellulose) that act as a substrate for the growth of beneficial microorganisms in the large bowel. **Probiotics** are beneficial living microorganisms (bacteria and fungi), usually consumed as part of fermented products such as yoghurt, that are claimed to provide added health benefits when ingested in adequate amounts.

Carotenoids are considered to be important dietary supplements: β-carotene acts as an antioxidant and is a precursor for vitamin A, lutein is believed to reduce the risk of macular degeneration, zeaxanthin is also associated with eye health while lycopene is linked to the support of a healthy prostate. Good sources of these nutraceuticals are carrots (carotene), broccoli (lutein), cooked tomatoes (lycopene) and goji berries (zeaxanthin).

Dietary fibres include four main groups of products with typical "soft claims" to promote their use. Insoluble fibres (wheat bran, corn bran and fruit skins) act as prebiotics and are considered to be important for digestive health; beta glucan (oat bran, oatmeal and rye) may reduce the risk of coronary heart disease; soluble fibres (e.g. psyllium seed husks, peas, beans, apples, citrus fruits, baobab) have the same claim but may also reduce the risk of some types of cancer; whole grains (cereal grains, whole wheat bread, oatmeal, brown rice) usually have the same claims as soluble fibres but also support the maintenance of healthy blood sugar levels.

Unsaturated fatty acids are believed to reduce the risk of coronary heart disease by lowering cholesterol. They include monounsaturated fatty acids (found in canola oil, olive oil and tree nuts) and polyunsaturated fatty acids (omega-3-fatty acids) found in flaxseed oil, flax seeds, walnuts and chia (seeds of *Salvia hispanica*; it has had novel food status in the EU since 2009).

Flavonoids are believed to be beneficial in acting as antioxidants and free-radical scavengers. These phenolic compounds include anthocyanins such as cyanidin, delphinidin, pelargonidin and malvidin (the pigments in berries, cherries and red grapes), flavonols and their derivatives such as catechins, epicatechins and epigallocatechins (in cocoa, chocolate, tea, apples and grapes), procyanidins and proanthocyanidins found in apples, grapes, red wine, chocolate, cinnamon, cranberries, peanuts and strawberries, flavanones such as hesperetin and naringenin in citrus fruits, and flavonols such as quercetin, kaempferol, isorhamnetin and myricetin in apples, broccoli, onions and tea.

Isothiocyanates (present in Brassicaceae such as broccoli, cabbage, cauliflower, horseradish and kale), plant stanols and sterols (present in pumpkin seeds and many types of oily seeds), phenolic acids (present in many fruits), minerals (e.g. calcium and magnesium) and vitamins are all included in the concept of dietary supplements. The classification of medicinal herbs as dietary supplements is a regulatory anomaly unique to the United States.

Examples of plants that are commonly promoted as health foods and dietary supplements are listed by their common names below.

acai berry (*Euterpe oleracea*)
African potato (*Hypoxis hemerocallidea*)
alfalfa – see lucerne
aloe vera (*Aloe vera*)
American spikenard (*Aralia racemosa*)

German postal stamp depicting Paracelsus

Ergot, growing on rye (Secale cereale)

Fly agaric (*Amanita muscaria*), one of the oldest hallucinogens

Strychnine (from *Strychnos nux-vomica*) was once a popular rodent poison

A list of the most deadly of all plant poisons is presented below. These are all classified in the WHO category Ia (extremely hazardous). Note that many of these species have close relatives that may be equally poisonous.

angel's trumpet (*Brugmansia suaveolens*); all parts: scopolamine, hyoscyamine, tigloidine, 3-tigloyloxytropane-6β-ol and other tropane alkaloids (up 0.4%); neurotoxin, mind-altering, medicinal plant

aspergillus (*Aspergillus flavus*); mycelia: aflatoxin; cell toxin, disturbance of GI tract, mutagenic

autumn crocus (*Colchicum autumnale*); all parts, especially seeds and bulbs: colchicine and related alkaloids; cell toxin, neurotoxin, medicinal plant, disturbance of the GI tract

bushman's poison (*Acokanthera oppositifolia*); stems, leaves, seeds: cardiac glycosides (ouabain); cell toxin, heart poison

calabar bean (*Physostigma venenosum*); seeds: physostigmine, eseramine, physovenine, and other indole alkaloids; neurotoxin, mind-altering, medicinal plant

castor oil plant (*Ricinus communis*); seeds: ricin (a mixture of four lectins), ricinine (pyridine alkaloid), ricinoleic acid (fatty acid); cell toxin, inflammatory, disturbance of the GI tract, medicinal plant

climbing lily – see flame lily

climbing potato (*Bowiea volubilis*); all parts, especially bulbs: scillaren-type bufadienolides: bovoside A, C, bowienine (alkaloid); cell toxin, heart poison, disturbance of GI tract, animal poison

cuckoo pint (*Arum maculatum*); aerial parts, fruits: aroin, cyanogenic glucosides, saponins, Ca^{2+}-oxalate crystals (sharp raphides which can penetrate cells); cell toxin, inflammatory, disturbance of GI tract

curare (*Strychnos toxifera*); all parts: toxiferine; neurotoxin, medicinal plant

curare vine (*Chondrodendron tomentosum*); aerial parts: curarine, tubocurarine; neurotoxin, mind-altering, medicinal plant

deadly nightshade (*Atropa belladonna*); all parts: hyoscyamine, scopolamine and other tropane alkaloids; neurotoxin, mind-altering, medicinal plant

dead-man's tree (*Synadenium cupulare*); all parts, latex: 12-O-tigloyl-4-deoxyphorbol-13-isobutyrate and several other tigliane-type diterpene esters of the 4-deoxyphorbol type; cell toxin, inflammatory, disturbance of the GI tract, animal poison

death camas (*Zigadenus brevibracteatus*); bulbs: zygadenine, zygacine, protoveratrine and related steroidal alkaloids; cell toxin, neurotoxin, mind-altering

death cap (*Amanita phalloides*); fruiting body: several peptides (amanitin and phalloidin); hallucinogen

desert rose (*Adenium obesum*); all parts: cardiac glycosides (obebioside); cell toxin, neurotoxin, heart poison

English yew (*Taxus baccata*); all parts (except red aril of fruits): taxin A, B, C; taxicin I, II; cell toxin, neurotoxin, medicinal plant

ergot (*Claviceps purpurea*); spore bodies (sclerotia): ergot alkaloids (ergotamine and others); neurotoxin, mind-altering

false morels (*Gyromitra esculenta*); fruiting body: monomethyl hydrazine; cell toxin, neurotoxin, disturbance of the GI tract

flame lily (*Gloriosa superba*); all parts, especially bulbs and seeds: colchicine, gloriosine, superbiine; cell toxin, neurotoxin, disturbance of the GI tract

fly agaric (*Amanita muscaria*); fruiting body: nitrogen compounds (ibotenic acid, muscimol, muscarine); neurotoxin, mind-altering

foxglove (*Digitalis purpurea*); all parts: several cardenolides (purpurea glycoside, lanatoside, digitoxin, digoxin); cell toxin, heart poison, disturbance of the GI tract, medicinal plant

gilled mushroom (*Lepiota helveola*); fruiting body: amanitins, phalloidins; cell toxin, neurotoxin, disturbance of the GI tract

green lily (*Schoenocaulon drummondii*); all parts, especially seeds: cevadine, veratridine, sabadine and other steroidal alkaloids; cell toxin, neurotoxin, mind-altering, disturbance of the GI tract

hellebore (*Helleborus viridis*); aerial parts: cardiac glycosides (bufadienolides), hellebrin, steroidal saponins (helleborin), ranunculoside, alkaloids (celliamine, sprintillamine); cell toxin, heart poison, disturbance of the GI tract, medicinal plant

henbane (*Hyoscyamus niger*); all parts, roots and seeds: hyoscyamine, atropine, scopolamine and other tropane alkaloids; neurotoxin, mind-altering, medicinal plant

jute mallow (*Corchorus olitorius*); seeds: cardenolides (corchorin, helveticoside, evonoside); cell toxin, heart poison, disturbance of the GI tract

karra (*Cleistanthus collinus*); all parts, leaves: heart poison, cleisthanin A, B (lignan glycoside); cell toxin, heart poison, disturbance of GI tract

larkspur (*Consolida regalis*); all parts, especially seeds: delcosine, lycoctonine and other terpene alkaloids; cell toxin, neurotoxin, mind-altering, animal poison

larkspur (*Delphinium elatum*); all parts, especially seeds: delphinine, nudicauline, staphisine, ajacine and other terpenoid alkaloids; cell toxin, neurotoxin, disturbance of the GI tract

lords and ladies – see cuckoo pint

mandrake (*Mandragora officinarum*); mainly roots: hyoscyamine, atropine, scopolamine and other tropane alkaloids; neurotoxin, mind-altering, disturbance of the GI tract, medicinal plant

mezereon (*Daphne mezereum*); all parts, especially red berries: mezerein (phorbol ester), daphnin (coumarin glycoside); cell toxin, inflammatory, disturbance of the GI tract

monkshood (*Aconitum napellus*); all parts (tubers): terpene alkaloids (aconitine); neurotoxin

naucleopsis (*Naucleopsis* species); latex: toxicariosides, cardenolides; cell toxin, neurotoxin, heart poison, disturbance of the GI tract

nux vomica (*Strychnos nux-vomica*); aerial parts, especially fruits and seeds: strychnine, brucine, colubrine and other monoterpene indole alkaloids; cell toxin, neurotoxin, mind-altering, medicinal plant

odollam tree – see suicide tree

oleander (*Nerium oleander*); all parts; nectar, even honey: oleandrin and several other cardenolides; cell toxin, heart poison, disturbance of the GI tract, medicinal plant

ordeal tree (*Erythrophleum suaveolens*); bark, aerial parts: cassaine, erythrophleine and other diterpenoid alkaloids; heart poison, disturbance of the GI tract, medicinal plant

Osage orange (*Maclura pomifera*); latex: taxicarioside, cardenolides; cell toxin, neurotoxin, heart poison, disturbance of the GI tract

poison hemlock (*Conium maculatum*); all parts, seeds: coniine, conhydrine and other piperidine

Curare poisons are traditionally used for South American blow darts

Bushman's poison (*Acokanthera oppositifolia*) – heart glycosides

Flame lily (*Gloriosa superba*) – colchicine

English yew (*Taxus baccata*) – diterpene pseudoalkaloids

alkaloids; neurotoxin, medicinal plant, mind-altering

poison leaf (*Dichapetalum cymosum*); aerial parts: monofluoroacetate is the main toxin; also dictamine and other furanoquinoline alkaloids, monoterpenes; cell toxin, mutagenic, neurotoxin, inflammatory, disturbance of the GI tract

poison olive (*Peddiea africana*); roots, aerial parts: Peddiea factor A1 and other diterpenoids (phorbolesters of the daphnane-type); cell toxin, neurotoxin, inflammatory, disturbance of the GI tract

purging croton (*Croton tiglium*); seeds: TPA, phorbol esters, crotonide (purine alkaloids), crotin (a toxic lectin); cell toxin, inflammatory, disturbance of the GI tract

rosary bean (*Abrus precatorius*); seeds: lectins (abrin A–D); cell toxin, neurotoxin

savin (*Juniperus sabina*); all parts, especially young twigs: 3–5% essential oil with sabinene and sabiylacetate, thujone, other monoterpenes; cell toxin, abortifacient, inflammatory, neurotoxin, mind-altering, animal poison

strophanthus (*Strophanthus gratus*); all parts, especially seeds: *k*-strophantoside, *k*-strophanthin, cymarin, strophantidol, periplocymarin and other cardenolides; cell toxin, neurotoxin, disturbance of the GI tract, heart poison, medicinal plant

suicide tree (*Cerbera odollam*); seeds: tanghin and other cardiac glycosides; cell toxin, heart poison, inflammatory

tamboti (*Spirostachys africana*); all parts, latex: diterpenes such as stachenone, stachenol; cell toxin, inflammatory, disturbance of the GI tract

thorn-apple (*Datura stramonium*); all parts, especially seeds and roots: hyoscyamine, scopolamine, atropine; neurotoxin, mind-altering, medicinal plant

upas tree (*Antiaris toxicaria*); aerial parts, latex: cardioactive glycosides (toxicariosides, antiarin derivates); cell toxin, heart poison

wild passion flower (*Adenia digitata*); all parts: lectin (modeccin), cyanogenic glucosides; cell toxin

witches tree (*Latua pubiflora*); all parts: hyoscyamine, atropine, scopolamine and other tropane alkaloids; neurotoxin, mind-altering

yellow heads (*Gnidia kraussiana*); all parts, fruits: phorbol esters of the daphnane type (kraussianin, gnidilatin, gnidilatidin), gnidicin, gnididin, gniditrin; cell toxin, inflammatory, neurotoxin, disturbance of the GI tract, animal poison

yellow jasmine/jessamine (*Gelsemium sempervirens*); all parts, roots, nectar: gelsemine, sempervirine and other indole alkaloids; cell toxin, neurotoxin, mind-altering

Plant parts used

The pharmaceutical names of plants and plant products are given in Latin in official documents and on product labels. This is not only to avoid confusion (with potentially fatal consequences), but also to make sure that the relevant plant part to be used is clearly specified. The chemical composition of roots, leaves, bark, fruits or seeds may be quite different, so that it is important to stipulate which part is suitable for use as medicine. Seeds, for example, may be quite toxic because of an accumulation of alkaloids, while the rest of the plant can be harmless in small doses. The latex from the fruit capsules of opium poppy (*Papaver somniferum*) is a source of opium alkaloids, while poppy seeds are practically alkaloid-free and used as a food and spice. (Be careful, however, as drug tests at airports may give a positive result even if you have only eaten a hamburger with poppy seeds on the bun!)

The following examples show how plant parts are indicated (in pharmaceutical names):

Entire plant (*herba tota*): the whole plant is used, roots and all (e.g. *Taraxacum officinale herba tota*).

Root (*rad.* or *radix*): fleshy or woody roots (or root bark) are used. Roots may be fibrous (stinging nettle, *Urtica dioica*; *Urticae radix*), solid (liquorice, *Glycyrrhiza glabra*; *Liquiritiae radix*) or fleshy (devil's claw, *Harpagophytum procumbens*; *Harpagophyti radix*).

Rhizome (*rhiz.* or *rhizoma*): a rhizome is a woody or fleshy elongated stem that usually grows horizontally below or at ground level, bearing leaves and roots. There is a clear distinction between roots and rhizomes yet the two are often confused. Examples of medicinal rhizomes include kava kava (*Piper methysticum*; *Kava-kava rhizoma*) and ginger (*Zingiber officinale*; *Zingiberis rhizoma*).

Bulb (*bulbus*): a bulb is a fleshy structure comprising several layers of fleshy leaf bases known as bulb scales. Examples are onion (*Allium cepa*; *Cepae bulbus*), garlic (*Allium sativum*; *Allii sativi bulbus*) and the European squill (*Drimia maritima*; *Scillae bulbus*).

Tuber (*tub.* or *tuber*): a tuber is a swollen, fleshy structure below the ground (often representing both stem and root). It is called a corm when a fibrous layer is present. Examples are hypoxis (*Hypoxis hemerocallidea*; *Hypoxidis tuber*) and autumn crocus (*Colchicum autumnale*; *Colchici tuber*).

Bark (*cort.* or *cortex*): bark is the outer protective layer of a tree trunk that is often periodically shed. It is formed from a layer of living cells (called the cambium) just above the wood itself. Chemical compounds often accumulate in bark (e.g. tannins and saponins), hence the reason why it is frequently used as medicine. Well-known medicinal barks include quinine (*Cinchona* species; *Chinae cortex*), pepperbark (*Warburgia salutaris*), oak bark (*Quercus* species; *Quercus cortex*) and willow bark (*Salix* species; *Salicis cortex*).

Wood (*lig.* or *lignum*): thick stems or the wood itself (often presented as wood chips) may be used as medicine. Examples include sandalwood (*Santalum album*; *Santali album lignum*) and quassia wood (*Quassia amara*; *Quassiae lignum*).

Leaf (*fol.* or *folium*): leaves alone may be used (*folium*), or leaves may occur in a mixture with petioles and twigs (*herba*). **Stems** (*stip.*, *stipes* or *stipites*) and even **stem tips** (*summ.* or *summitates*) are sometimes specified. Examples are the maidenhair tree (*Ginkgo biloba*; *Ginkgo folium*) where only the leaves are used, and bittersweet (*Solanum dulcamara*; *Dulcamarae stipites* or *stipes*) where the leafless stems (two or three years old) are used.

Aerial parts (*herba*): all aboveground parts are harvested as medicine, often just before or during flowering. Examples include spilanthes (*Spilanthis oleraceae herba*) and St John's wort (*Hypericum perforatum*; *Hyperici herba*).

Flowers (*flos*): flowers are sometimes used, such as cloves (the flower buds of *Syzygium aromaticum*; *Caryophylli flos*), chamomile flowers (*Matricaria chamomilla*; *Matricariae flos*) and Roman chamomile flowers (*Chamaemelum nobile*; *Chamomillae romanae flos*). Particular flower parts may be specified, such as hibiscus calyces (*Hibiscus sabdariffa*; *Hibisci flos*), the style branches of saffron (*Crocus sativus*; *Croci stigma*), the stigmas ("beard") of maize (*Zea mays*; *Maidis stigmas*) or even pollen (*pollinae*). The whole inflorescence or young infrutescence is sometimes used, such as the "cones" of hops (*Humulus lupulus*; *Lupuli strobulus*).

Fruit (*fr.* or *fructus*): fruits vary in structure and are sometimes wrongly referred to as seeds. An example is the small dry schizocarps in the Apiaceae

Examples of plant parts used in herbal medicine: roots (liquorice), rhizomes (ginger), bark (oak), wood (quassia), leaves (ginkgo), all aerial parts (St John's wort), flowers (chamomile), fleshy fruits (rose hips), dry fruits (cumin) and resin (myrrh)

family, each comprising two one-seeded mericarps that usually split apart at maturity. Examples are fennel fruits (*Foeniculum vulgare*; *Foeniculi fructus*) and anise (*Pimpinella anisum*; *Anisi fructus*). Small dry one-seeded nutlets (achenes) may be used, such as milk-thistle achenes (*Silybum marianum*; *Cardui mariae fructus*) or fleshy fruits or cones may be specially dried, such as saw palmetto fruits (*Serenoa repens*; *Sabal fructus*) or juniper "berries" (*Juniperus communis*; *Juniperi fructus*). Only specified parts of the fruit may be suitable, such as edible rose hip pericarps (*Rosae pericarpium*), or inedible pomegranate peel (*Punica granatum*; *Granati pericarpium*) and bitter-orange peel (*Citrus aurantium*; *Aurantii pericarpium*).

Seed (*sem.* or *semen*): seeds are contained within a fruit and may be used with or without the fruit pericarp. Examples include the true seeds (nuts) of the castor oil plant (*Ricinus communis*; *Ricini semen*) and fenugreek seeds (*Trigonella foenum-graecum*; *Foenugraeci semen*).

Gum (*gummi*): gums are solids consisting of water-soluble mixtures of polysaccharides. Gum may function as a defence mechanism to stop wood-boring insects and to seal off wounds so that wood-rotting fungi and bacteria are kept out. An example of an exudate gum is gum arabic (from *Acacia senegal*; *Gummi acaciae*) that is still used in the pharmaceutical industry. Gums mixed with water are known as gels. An example is the gel present in the inner leaf pulp of *Aloe ferox* and *A. vera*.

Resins (*resina*): plants have specialised ducts, glands or cells that excrete resins. Resins are mixtures of essential oils and polymerised terpenes. Unlike gums, they are usually insoluble in water.

Examples are frankincense (from *Boswellia sacra*; *Olibanum*), myrrh (from *Commiphora myrrha*; *Myrrha*) and mastic (*Pistacia lentiscus*; *Resina mastix* – used as an adhesive for dental caps). Balsams (or balsamic resins) are resins with a high content of benzoic acid, cinnamic acid or their esters. Well-known examples include Tolu balsam (from *Myroxylon balsamum* var. *balsamum*), Siam benzoin (from *Styrax tonkinensis*) and Sumatra benzoin (*Styrax benzoin*). Storax balsams are collected from *Liquidambar* species (do not confuse them with balsams from *Styrax* species): Levant storax ("balm of Gilead" in the Bible) comes from *Liquidambar orientalis*; common storax from the sweet gum tree (*L. styraciflua*).

Fatty oil (*oleum*): non-volatile vegetable oils, insoluble in water, that are found in seeds or fruits. Oils are described as acylglycerides because they are formed from a glycerol molecule that is attached to various types of fatty acids. Castor oil (from *Ricinus communis* seeds) is an example with direct medicinal (laxative) properties, while others (almond oil, olive oil, safflower oil) are used as carrier oils in liquid formulations and ointments (e.g. in aromatherapy).

Essential oil (*aetheroleum*): these are volatile oils (= essential oils), selectively obtained from plants through steam distillation or through extraction with a non-polar solvent such as hexane. They consist mainly of monoterpenoids, sesquiterpenoids, phenylpropanoids and coumarins, and are important as biologically active ingredients of plants. Examples are camphor (from the wood of *Cinnamomum camphora*; *Camphorae aetheroleum*) and peppermint oil (from leaves of *Mentha* ×*piperita*; *Menthae piperitae aetheroleum*).

Dosage forms

Initially, phytomedicines were simply eaten, chewed, snuffed or applied externally; more sophisticated dosage forms such as infusions and tinctures developed later. The traditional methods of preparation were often aimed at eliminating some toxins or increasing the efficacy. The volume of plant materials used in relation to the volume and type of solvent used are critical. Plants are typically very variable but experienced traditional healers knew how to adjust dosages after an appraisal of the effects that the medicine had on their patients. Today, phytomedicines are carefully manufactured and standardised to ensure safety and efficacy. Dried leaves, roots and other plant parts are still widely used in the form of tea but specialised extracts and tablets are also available. Drug delivery systems of the future may involve microtechnology and nanotechnology. Sometimes the extract is manipulated to increase the concentration of desired compounds while eliminating or reducing unwanted substances. Since plants cannot be patented, such special extracts are often registered, branded and sold as proprietary products. There is a tendency to disregard traditional dosage forms (such as the old-fashioned tea) in favour of modern galenic dosage forms such as tablets and pills but the former is often much more appropriate and cost-effective.

Various dosage forms are listed alphabetically below.

Capsule: a small container (usually two gelatin halves sliding over one another) that contains powdered medicinal products or extracts in an exact dose. The content is protected from moisture, light and air. The capsule wall contains a softening agent such as glycerol or sorbitol to ensure that it readily dissolves. It is made from gelatin (an animal product) but special vegecaps (Vcaps™) are available for vegetarians. Soft gelatin capsules are spherical or ovoid containers used for oily (water-free) extracts, in liquid or semisolid form.

Decoction (*decoctum*): an extract prepared by adding cold water to the crude dug and then boiling it for 5–10 minutes (sometimes longer, for several hours). The concentration of some substances may be much higher than in an infusion prepared with the same weight and volume of starting materials.

Extract (*extractum*): a mixture of soluble chemical compounds, separated from the unwanted fibrous and non-soluble portion of a crude drug using water or alcohol (ethanol). An extract may be liquid or viscous but is often dried and powdered. Volatile oil can be extracted in hexane but is more often separated by steam distillation. The herb to extract ratio (HER) is typically 5:1 for normal extracts, or about 100:1 for essential oils. A modern method is supercritical fluid extraction, using liquid carbon dioxide. Liquid extracts are usually prepared in such a way that one part by volume of the preparation is equivalent to one part by weight of the crude drug. Soft extracts are prepared by evaporating the solvent until a soft mass is produced.

Granules: small particles produced by combining a concentrated powdered extract with a soluble excipient such as gelatin, lactose or sucrose. The granules are usually included in capsules or pressed to form tablets.

Infusion (*infusum*): an extract (usually referred to as "tea") prepared by adding boiling water to the crude drug and allowing it to steep for 5–10 minutes (without boiling). Infusions are easily contaminated by microbial growth and should be used within 12 hours.

Inhalation: a liquid preparation with volatile substances that are inhaled in order to treat the lining of the respiratory tract (nose, throat and lungs). The active ingredients may be volatile at room temperature or they are heated with hot (65 °C, not boiling) water and the vapour inhaled for 5–10 minutes.

Instant tea: a dried herbal extract mixed with a suitable filler or carrier. The carrier (typically lactose, sucrose or maltodextrin) increases bulk, reduces viscosity and improves solubility. Spray-drying is mostly used, where a concentrated infusion of the herb is sprayed at high pressure (as a mist) into a heated column with suspended particles of the carrier, which become coated with the herbal extract as it dries.

Juice (*succus*): a liquid prepared by crushing fresh plant material in water and then expressing the juice. Commercial juice is pasteurised or treated with ultra-high temperature to extend the shelf life.

Linctus: a viscous liquid preparation, usually containing sugar and medicinal substances and used for its demulcent, expectorant or sedative proper-

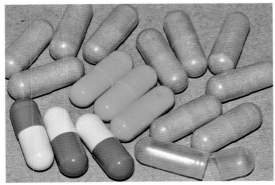
Gelatin capsules

Soft gelatin capsules

Infusion and decoction

Lozenges

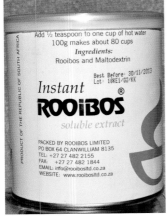
Instant tea

Juices

Linctus

ties. They are sipped and swallowed slowly without the addition of water.

Liniment: a liquid or semi-liquid preparation intended for external application. It may contain substances with analgesic, rubefacient, soothing or stimulating activities.

Lotion: an aqueous or alcoholic solution or watery suspension intended for application to the skin.

Lozenge (pastille): neatly shaped lumps of moulded and solidified sugar containing medicinal extracts and intended for sucking or chewing, so that the active ingredients are slowly released in the mouth. They often contain smaller quantities of gums, colourants and flavourants.

Maceration (*maceratio*): an extract prepared by adding cold water to the crude drug and allowing it to soak at room temperature for 6–8 hours.

Medicinal oil: a medicinal extract dissolved in fatty oil or liquid wax and intended for internal or external use (e.g. garlic oil). Oils used in aromatherapy comprise an essential oil (or mixture of oils) dissolved in a carrier oil (often almond oil) or liquid wax (jojoba).

Medicinal spirits or medicinal essence: volatile compounds (usually essentials oils) dissolved in alcohol or alcohol–water mixtures. It may be produced by mixing aromatic herbs with alcohol and then recovering a mixture of the volatile compounds and the alcohol by steam distillation.

Mixture: two or more medicinal herbs that are used in a fixed combination. Mixtures are typical in Traditional Chinese Medicine and African Traditional Medicine, where the individual components may act individually and additively or even synergistically to restore or maintain health.

Monopreparation: a medicinal product that contains only a single herb (or extract) as active ingredient.

Nasal drops: liquid preparations instilled into the nostrils by means of a pipette.

Ointment, paste and gel: semi-solid preparations of medicinal substances dissolved in watery and/or oily solvents or emulsions and intended for external application.

Pill: a neatly shaped, small solidified lump made from a semisolid mixture of medicinal substances with a suitable bulking agent such as gum arabic. The material is cut into small portions of equal size and weight, rolled or shaped and then allowed to solidify. Pill-making was once part of the practical training of a pharmacist but the process is nowadays fully mechanised.

Snuff: dried and finely powdered plant material that can be drawn up into the nostrils through inhalation. Sneezing is traditionally thought of as a way to expel an ailment.

Special extract: a mixture of desirable chemical compounds, obtained by modifying and manipulating the extraction solvents (e.g. liquid carbon dioxide) in such a way that unwanted substances are reduced or eliminated.

Suppository: an oblong, tablet-like product that is inserted into the rectum, vagina or urethra and left to dissolve there. Herbs are more commonly used as enemas (liquid or semisolid preparations).

Syrup (*sirupus*): a viscous sugar solution used as flavouring or taste-masking agent or as cough medicine. It is usually a saturated solution of sucrose (66%) but should not contain less than 50%. Syrups are sterile because there is no free water available for microbial growth. When used as a cough remedy (*linctus*), the syrup is sipped slowly to ensure maximum contact with the inflamed mucous membranes.

Tablet: a neatly shaped, small lump made by compressing a mixture of powdered active ingredient and an excipient (inert binder and bulking agent). A colourant, flavourant and disintegrator may be added (the last-mentioned to ensure that the tablet rapidly dissolves when placed in water). Coated tablets are covered in a thin layer of sugar, colouring agent, fat, wax or special film-forming agents to improve their appearance, to mask an unpleasant taste, to make them smooth for easier swallowing, to prevent them from dissolving too rapidly and to improve shelf life. Film-coated tablets have a surface layer of cellulose acetate phthalate or other substances that resist gastric juices. This is to protect the stomach lining or to ensure that the active substances are only released when the tablet has reached the bowel.

Tea bag: a small porous paper container with a fixed quantity (dose) of finely chopped plant material. They are convenient to use but have a relatively short shelf life because the large surface area promotes not only rapid extraction but also the disadvantages of oxidation by air and evaporation of volatile compounds.

Tea mixture (*species*): a fixed mixture of herbs used for a specific indication (usually four to seven, included for specific purposes). There are active herbs (containing chemical compounds known to be of pharmaceutical benefit for the specific indication), supplementary herbs (supportive of the indication) and adjunct herbs (added to improve the taste, smell or colour of the mixture). Examples are *species amaricantes* (= bitter tea), *species anticystiticae* (= bladder tea), *species carminativae* (= carminative tea), *species laxantes* (= laxative tea) and *species sedativae* (= sedative tea or nerve tea).

Tea: an infusion made covering the herb in boiling water and allowing it to steep for several minutes while the water cools down. The word "tea" when used alone usually refers to black tea (*Camellia sinensis*), known in many parts of the world as *chai* (after the Cantonese and Mandarin name). In

Pills

Tablets

Tea bags

Tea mixture

African teas

African tinctures

the case of other "teas" the raw material has to be specified, e.g. hibiscus tea (*Hibiscus sabdariffa*) or rooibos tea (*Aspalathus linearis*). Sometimes the word "tea" may be ambiguous. Ginger tea or mint tea, for example, may refer to tea made from ginger or mint or to black tea flavoured with ginger or mint. Teas may be taken as pleasant hot drinks or for their therapeutic benefits and the distinction is not always clear.

Tincture (*tinctura*): an extract prepared by soaking the crude herb in an alcoholic solution (usually 70% alcohol) for a specified period, after which it is pressed and strained. The original extract (mother tincture) may then be diluted with pure water to a predetermined herb to extract ratio. No preservatives are necessary because the alcohol prevents microbial growth. To avoid alcohol, glycerides may be prepared by using glycerol as the solvent instead of alcohol.

Methods and routes of administration

The application of a medicinal substance may be topical (local) if it is applied on a localised part of the body, enteral when it is given via the digestive tract and parenteral when it is directly injected into the body. Topical may also refer to epicutaneous (applied on the skin), inhalational (applied into the lungs, through inhaling or smoking), enematic (applied into the rectum, as suppository), conjunctival (applied to the eye), otic (applied into the ear) and mucosal (applied to mucous membranes of the nose (through insufflation – snuffing, snorting); tongue (sublingual); between the lips and gums (sublabial); vaginal or rectal routes of administration. Enteral is application by mouth (or gastric feeding tube). Parenteral application includes intravenous, intraarterial, intramuscular and subcutaneous. The route of administration is critical because some substances may be highly toxic when injected but harmless when ingested.

Bathing: herbs and herbal mixtures may be added to bath water to alleviate pain, to treat skin conditions and to simply maintain good health.

Conjunctival and otic application: medicinal substances are applied in the form of eye drops and ear drops. These preparations usually have anti-inflammatory, analgesic and antiseptic activities.

Ingestion: infusions (teas), decoctions, syrups and tinctures are usually taken orally (by mouth) and swallowed. The medicinal substances are subjected to acid hydrolysis in the stomach; some compounds may be lost, while others may be converted to the active form. In many cases, the compounds taken by mouth are merely the prodrugs. They may be converted into the active drugs in the stomach (through acid hydrolysis) or in the colon (through the action of bacterial enzymes). Some compounds may be inactivated in the stomach or may be metabolised in the liver.

Injection (parenteral application): suitable preparations are introduced directly into the bloodstream, usually with hypodermic needles and syringes. There are various other methods used in traditional societies, such as rubbing substances into small breaks or cuts in the skin (e.g., the Zulu practice of *umgaba*). Injection is a highly effective but also potentially lethal method. Many substances that are harmless when ingested can be deadly when injected (oral toxicity is often an order of magnitude less than parenteral toxicity).

Mastication: some herbs are typically used as masticatories and are held in various parts of the mouth (sublingual, sublabial), sometimes for extended periods. Active compounds are directly absorbed into the rich supply of blood vessels below the mucous membranes of the mouth and may rapidly reach the brain, unaltered by stomach acids or the liver. It is likely that humans once took all or most of their medicine in this way. Examples include chewing tobacco, betel nut, hoodia and sceletium. In traditional medicine systems, many herbs are used as snuff, not only to treat headache but also for the belief that the induction of sneezing will help to expel the disease. Powdered tobacco (snuff) and cocaine are typically used in this way.

Rectal or vaginal application: infusions and decoctions may be administered as enemas, using modern enema syringes or tubes. Specially prepared suppositories can also be used, so that the active ingredients are slowly released and absorbed over an extended period of time. The popularity of enemas varies greatly among healing cultures and different societies.

Rinsing and gargling: mouth rinses and gargles are usually aimed at antimicrobial (antiseptic) activity and to soothe infected and inflamed mucous membranes of the mouth and throat. They are also used for oral hygiene and the prevention of plaque and dental decay (or simply to freshen the breath). The products used are typically not swallowed, as they are often potentially toxic if ingested.

Smoking: smoking is a popular method to treat asthma and other respiratory conditions but is better known as the most popular method for inducing mind-altering and sedative effects. Many plant products have been used in this way, the most famous being tobacco, cannabis, opium and thorn apple (*Datura*). Smoking is sometimes used in traditional medicine to induce coughing, which is considered to be a way of expelling the perceived cause of the ailment. Various parts of the body may be subjected to smoke treatment or fumigation, for pain relief and to soothe the skin.

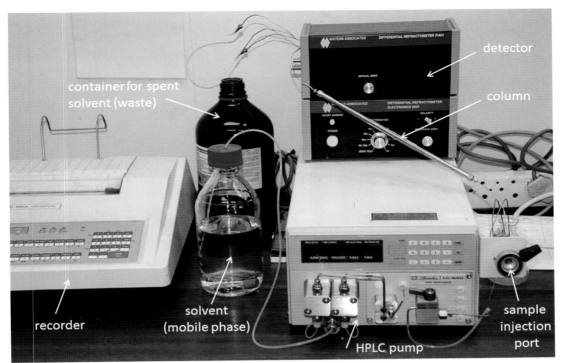

High-performance liquid chromatography (HPLC) system

Identification and structural determination is the main routine task performed by phytochemists. The identity of isolated (pure) compounds can be determined using a combination of chromatography and other methods such as UV-vis spectroscopy, infrared spectroscopy (IR), mass spectrometry (MS) and nuclear magnetic resonance spectroscopy (NMR). In spectroscopy, the absorption and reflection of visible, UV and infrared light is measured and compared to the values or patterns for known compounds and reference standards. Mass spectrometry is a method in which the compound is broken down in a highly controlled way and the fragments are measured and compared. Each compound has a characteristic fragmentation pattern that serves as a "fingerprint", unique for all except stereoisomers. The ultimate method to determine the absolute configuration of a molecule is by NMR, a sophisticated technique in which the pure compound is placed within a strong magnet and the unique pattern of interactions (couplings) between atoms within the molecule is measured and analysed by organic chemists to determine the structure. In addition, the optical rotation of the compound can be measured (useful in case of optical isomers). When crystals are available, the absolute configuration can be determined by crystallographical methods. The melting point of crystals can provide further confirmation of identity.

A modern approach to studying the chemical compositions of plant extracts is known as metabolomics. It uses LC-MS and NMR techniques with sophisticated computer software to study the total metabolome of an organism (animals or plants). All of the thousands of primary and secondary metabolites are analysed at the same time in an attempt to understand the biology and physiology of the organism. The method is often used in diagnostics and toxicology. For example, blood samples taken before and after a compound or medicine was administered can be used in comparative studies of pharmacodynamics and to see what main effects the test compound had on the metabolism of the animal or person. Principal components analysis (PCA) is a popular statistical method in this type of work and is often performed to determine the main phytochemicals responsible for a change in the metabolome. Another modern development is the use of DNA fingerprinting in the authentication of herbal medicines.

Although highly sophisticated methods such as GC-MS, LC-MS and metabolomics are now available for routine work, there remains a need for "old-fashioned" chromatography such as TLC, GLC and HPLC. These methods are still being used as part of the daily routine in the food, cosmetics and pharmaceutical industries for quality control purposes and in research and forensic laboratories.

Quality control and safety

The safety and efficacy of phytomedicines and natural substances are ensured through the process of quality control. This is done by double-checking not only the correct identity of the crude herb or extract, but also the required concentration (dose) of active ingredients and the purity of the product (to confirm the absence of adulterants, undesirable chemicals or biological contaminants such as bacteria). All procedures are carefully documented so that every step in the production process can later be verified in case side effects or other unexpected outcomes are reported. The focus of quality control procedures is more on safety and quality than on efficacy. The latter is based on traditional evidence and experience and in some cases also on controlled clinical trials.

Pharmacognosy is the science that deals with the identification of medicinal plants and drugs. It is very important to ensure that toxic plants are not mistaken for the required species. The process requires botanical knowledge (to identify the correct plant or its parts), anatomical knowledge (to identify characteristic tissues or cells, such as glands) and phytochemical knowledge (to identify the main chemical compounds or to compare a chromatographic fingerprint with that of an authentic reference sample). Species identification can also be achieved by DNA barcoding.

Purity and hygiene are essential requirements for both the raw materials and the finished products. They have to be free from adulterants, because the adulterant may either be toxic or it may dilute the product so that it is no longer effective. Foreign organic matter may include bits of unwanted plant materials from the same species (e.g. leaf material when only flowers are harvested) but more often material from related or even unrelated species that were intentionally or inadvertently included in the product. In the latter case it may not exceed 2%. Soil and inorganic contaminants can be introduced when herbs are air-dried in the open under windy and dusty conditions, or soil may adhere to rhizomes and roots if they are not carefully washed before drying and milling. This form of contamination can be monitored by the ash value of the material: the % weight of the ash that remains after a sample of the product has been incinerated. The total ash value that is acceptable is often fixed at 3% or 5% of dry weight, and acid-soluble ash at no more than 1%, depending on the product. Other important variables that need to be controlled are bacterial contamination, heavy metals, organic pesticides and radioactive residues.

Microbial contaminants are controlled as a matter of routine through standard microbiological tests. The maximum acceptable levels of microorganisms that are allowed depend on the intended use of the product. Material intended to be used for teas and tinctures, for example, will be sterilised by the heat or alcohol when they are prepared for use, so the requirements for them would be less stringent than for an ointment that is applied to open wounds. *Salmonella* species and *Escherichia coli* must usually be negative (not detectable), while the upper limit is usually 10^4 per gram for fungi and 10^5 per gram for aerobic bacteria.

In the final dosage form, heavy metals may not exceed 10 mg/kg (10 parts per million) in the case of lead and 0.3 mg/kg for cadmium. Pesticide residues are monitored according to international guidelines. The upper limit for aldrin and dieldrin is usually set at 0.05 mg/kg. International guidelines also apply to the levels of radioactive residues in plant materials, such as strontium-90, iodine-131, caesium-134, caesium-137 and plutonium-239.

Standardisation is the process through which a finished product (phytomedicine) is manufactured in such a way that all batches contain the same amount of the active chemical compounds (or carefully selected marker compounds). The activity of herbal medicines is often not linked to a single chemical entity, so that a convenient marker is chosen as a standard. It is assumed that the active compound(s) are present at therapeutic levels if the marker compound is above a specified level. The concentration or dosage is very important because the active ingredients may produce serious side effects (if the dose is too high) or have no therapeutic value at all (if the level is too low).

Most medicinal plants have a very wide therapeutic window, meaning that the therapeutic dose is much lower than the toxic dose. For this reason, a minimum dose level is usually specified for medicinal herbs, rather than an upper limit. Plant material is often highly variable, so that it is difficult to set an exact level, as is done with pure chemical substances. An upper limit is, however,

Comparison of rooibos tea samples

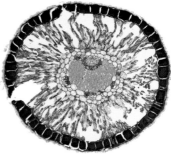
Transverse section through the needle-shaped leaf of rooibos tea

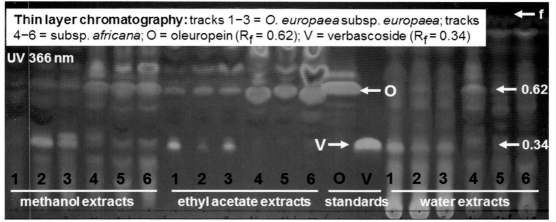
Quality control of olive leaf extracts

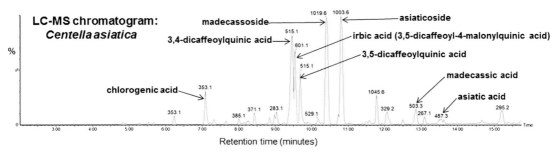
LC-MS chromatogram of *Centella asiatica* showing the main active compounds

essential for plants containing chemicals with a narrow therapeutic window, such as alkaloids and cardiac glycosides. These products can usually only be taken on prescription and under the supervision of a health care professional.

The safety of herbal medicines is an important consideration. Safety studies, performed according to strict protocols, are required before a new phytomedicine can be registered. Traditional medicines usually have a long history of safe use, so that safety is generally accepted on the basis that no serious side effects have been reported in the literature. It is possible that chronic toxicity may be overlooked (e.g. the cumulative effect of herbs such as comfrey and borage that contain pyrrolizidine alkaloids with an unsaturated necine base). Safety studies often involved test animals but such methods have become less popular. Tissue culture systems can also be used as a first assessment of toxicity. The safety and absence of serious side effects are usually the first parameters that are examined in a formal clinical trial.

Efficacy of herbal medicine

Efficacy is a critical consideration: does the product actually give the desired health benefit that it is intended to do? Many people are still sceptical about traditional medicine despite the fact that it has been used on a daily basis by many cultures all over the world for thousands of years. Furthermore, miracle cures such as aspirin for headache, artemisinin for malaria and vincristine for leukaemia leave no doubt that plants do have definite therapeutic value. However, unrealistic claims on product labels and marketing materials, as well as poor quality products containing little or no active ingredient are the main reasons why there is considerable and justifiable scepticism about efficacy.

The role of the placebo effect is often under-estimated. Many products rely on this powerful force within the human mind and body that regulates health and healing. The uniforms worn by traditional and modern health care workers, the way in which they make a convincing diagnosis and even the way in which medicinal products are formulated and packaged may all contribute towards strengthening the placebo effect. Traditional healers often rely on clever interventions, such as adding powdered aspirin or a mild laxative to the medicine they prescribe, so that there are detectable physical effects in the patient that will serve to enhance the placebo effect (i.e., the belief that "the medicine is working"). The ultimate scientific test is the double-blinded, placebo-controlled clinical trial, aimed at objectively distinguishing between real therapeutic activity and placebo.

There are several ways in which the efficacy of traditional medicine can be assessed, ranging from the traditional use as the first level, to double-blinded placebo-controlled clinical trials and finally the blind testing against an established drug as the ultimate and most sophisticated level. Traditional use may often provide the first hypothesis, especially when a particular medicine has been used for a particular ailment over many years. Studies of the chemical constituents of a plant may provide a scientific rationale for efficacy that has been determined empirically through regular use over many generations.

Plausible anecdotes (clinical observations by health care practitioners) are another source of evidence that may support the claimed efficacy of a traditional medicine. Several anecdotes, reported by independent observers, make the claim even more plausible, especially if it is backed up by physical evidence such as blood tests and pathology reports.

Pharmacological studies are done to demonstrate pharmacological activity using *in vitro, in vivo* and *ex vivo* methods. Such preclinical studies only give a superficial assessment, because isolated cells rarely behave in the same way as they would do when they are part of a complicated organ system that is connected to other systems in the human body. Plants often contain complex mixtures of chemical compounds and these may have unpredictable synergistic effects on various organ systems that cannot be determined by testing the compounds individually. Extracts or compounds may show activity when applied directly to cells in tissue culture but this is far removed from the real situation in the body, where the compound may not be absorbed or may never reach the cells where it is needed. For this reason, a study of pharmacokinetics is important, in order to determine how the compound is absorbed, circulated and eliminated.

Observational studies are performed when many people have already used the medicine out of their own free will for an extended period of time. If sufficient numbers of people are available, statistical methods can be applied to show significant health benefits in those who have used the medicine compared to those who have not (the control group).

Clinical studies (studies in humans) are the most convincing scientific method of proving safety and efficacy. Unfortunately, there are not many clinical studies done on herbal medicine because there is no way to protect the intellectual property, so that those who have funded this often very expensive activity will never get a return on their investment. Plants cannot be patented, and the results of a clinical study will place the competitors who did not fund it at an unfair advantage. One common approach to partly overcome this problem is to develop a patented proprietary extract, and to do the clinical studies on the extract and not the plant. The development of new medicinal products is done in four phases (a preclinical phase and three clinical phases). Phase I

 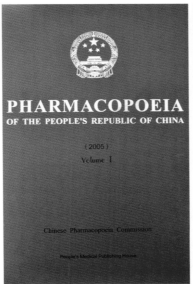 

Examples of pharmacopoeias

In Japan, herbal products are classified according to how they are regulated, either as prescription drugs, generics or over-the-counter drugs (*Pharmaceutical Affairs Act*) or as functional foods (*Nutritional Improvement Act*). They are also classified according to how they are marketed. *Kampo* drugs are mixtures of between two and 32 different species, formulated strictly according to ancient Chinese compendia. These are legally considered to be both prescription drugs and generic drugs – 130 such formulas are listed in the National Health Insurance Drug List. Herbs in Japanese pharmacopoeia are excluded because they are used alone and not in mixtures. Over-the-counter herbal drugs are based on traditional family formulas or old Chinese compendia. There are 210 such formulas (dosage and therapeutic claims are limited). The registration route for new drugs is expensive and applies to any new synthetic or plant-derived product. Functional foods are regulated in much the same way as generic drugs, and limited claims are allowed. Health foods are becoming very popular, but no therapeutic claims are allowed. The 16th edition of the Japanese Pharmacopoeia was published online and a new 17th edition is planned for 2016.

In Australia they use a relatively simple procedure, the so-called *listing system*. Herbal medicine is regulated by the *Therapeutic Goods Act* of 1989, which is aimed at controlling the quality, safety, efficacy and timely availability of therapeutic goods. Unless specifically exempted, all therapeutic goods must be entered on the *Australian Register of Therapeutic Goods* (ARTG) before they can be marketed, imported or exported. Herbal medicine may either be registered (as prescription drugs or as over-the-counter drugs) or more often listed in the register. The listing system also has a list of Australian Approved Names (AANs) for all ingredients made from plants or plant-like materials.

In China, safety or clinical data are required for new medicines, while traditional herbal medicines are listed in the *Pharmacopoeia of the People's Republic of China*. The latest English edition was published in two volumes in 2007. Volume 1 contains herbal medicine and Volume 2 Western medicine. It is the ninth edition since the first one was published in 1930.

In India, any herbal medicine can be registered as long as it appears in standard texts, as part of the Ayurvedic or Unani medicine systems. The official Indian Pharmacopoeia was published in four volumes by the Indian Pharmacopoeia Commission late in 2013. It came into effect on 1 January 2014 and features 30 additional new herbal drug monographs in Volume 3.

Aleppo rue (*Ruta chalepensis*): essential oil, flavonoids

Shan shu yu (*Cornus officinalis*): iridoids, triterpenoids

Siberian ginseng (*Eleutherococcus senticosus*): triterpene saponins

Soapbark tree (*Quillaja saponaria*): saponins

Poison rope (*Strophanthus speciosus*): cardiac glycosides

Common fig (*Ficus carica*): furanocoumarins

Raspberry (*Rubus idaeus*): tannins

Deadly nightshade (*Atropa belladonna*): tropane alkaloids

Secondary metabolites

The biological and pharmacological activities of medicinal, poisonous or mind-altering plants can be explained through the presence of mixtures of secondary metabolites often belonging to several structural classes. In many instances, individual compounds not only exert additive but also synergistic interactions. Secondary metabolites have apparently evolved as a means for plants to defend themselves against herbivores and against bacteria, fungi and viruses. They also serve as signal compounds to attract pollinating and seed-dispersing animals, furthermore as antioxidants and UV protectants. It is now generally accepted that it may not always be possible to solve complicated health problems with a single chemical compound or "magic bullet". It is far more likely that the sophisticated medicines of the future will be based on combinations of chemical compounds, targeting different organ systems and enzymes, wherever an imbalance needs to be rectified. From a biosynthetic perspective we can group secondary metabolites into those without nitrogen or those with nitrogen in their structures. Each group is again divided more or less according to their biosynthesis and chemical structure, resulting in eight main groups or classes as shown below.

A. Nitrogen-free compounds

1. Terpenes (p. 50)

Monoterpenes (p. 50)
Iridoid glucosides (p. 52)
Sesquiterpenes and sesquiterpene lactones (p. 52–55)
Diterpenes (p. 56)
Triterpenes and steroids (p. 58)
Saponins (p. 60)
Cardiac glycosides (p. 62)
Tetraterpenes and polyterpenes (p. 63)

2. Phenolics (p. 64)

Phenylpropanoids (p. 64)
Diarylheptanoids (p. 66)
Coumarins and furanocoumarins (p. 66)
Lignans and lignin (p. 67)
Flavonoids, stilbenoids, chalcones and anthocyanins (p. 67)
Catechins and tannins (p. 69)
Small reactive molecules (ranunculin, tuliposide, ethanol) (p. 70)

3. Quinones (p. 72)

Quinones and naphthoquinones (p. 72)
Anthraquinones and other polyketides (p. 72)

4. Polyacetylenes, polyenes, alkamides (p. 73)

5. Carbohydrates (p. 74)

6. Organic acids (p. 76)

B. Nitrogen-containing compounds

7. Alkaloids (including amines) (p. 78)

Amaryllidaceae alkaloids (p. 78)
Bufotenin, tryptamines and tyramines (p. 78)
Colchicine (p. 79)
Diterpene alkaloids (p. 79)
Ergot alkaloids (p. 80)
Indole alkaloids (including monoterpene indole alkaloids) (p. 81)
Indolizidine alkaloids (p. 82)
Isoquinoline alkaloids (including protoberberine, aporphine, morphinane alkaloids) (p. 82)
Phenylpropylamines (p. 84)
Piperidine alkaloids (p. 84)
Purine alkaloids (p. 84)
Pyrrolidine alkaloids (p. 85)
Pyrrolizidine alkaloids (p. 85)
Quinolizidine alkaloids (p. 85)
Quinoline alkaloids (including acridone alkaloids) (p. 85)
Steroidal alkaloids (p. 86)
Tropane alkaloids (p. 86)

8. Amino acids and related compounds (p. 88)

Amino acids (p. 88)
Non-protein amino acids (p. 89)
Cyanogenic glucosides (p. 90)
Glucosinolates and mustard oils (p. 90)
Lectins and peptides (p. 91)

Terpenes

Terpenes are built from C5-units as a building block and can be subdivided into hemiterpenes (C5), monoterpenes (C10), sesquiterpenes (C15), diterpenes (C20), triterpenes (C30), tetraterpenes (C40) and polyterpenes. Steroids (C27) are derived from triterpenes. Hemiterpenes are quite rare. An example is tuliposide A, one of the skin irritant compounds in tulip bulbs (see below). Most of the terpenoids are lipophilic. They readily interact with biomembranes and membrane proteins. They can increase the fluidity and permeability of the membranes, which can lead to uncontrolled efflux of ions and metabolites and even to cell leakage, resulting in cell death. In addition, they can modulate the activity of membrane proteins and receptors or ion channels. This membrane activity is rather non-specific; therefore, terpenes show cytotoxic activities against a wide range of organisms, ranging from bacteria and fungi to insects and vertebrates. Many terpenes are even effective against membrane-enclosed viruses.

Monoterpenes

Monoterpenes are derived from the head-to-tail coupling of two isoprene units (2 × C5). More than 40 skeleton types are known. Monoterpenoids occur in nature as the major components of volatile oils (essential oils), often in combination with sesquiterpenoids. Volatile phenolic compounds may also be present in essential oil, such as the phenylpropanoids eugenol and anethole (see under phenolic compounds). The oil is stored in specialised canals, resin ducts and trichomes (unicellular or more often multicellular glands) and is usually extracted by steam distillation.

Monoterpenes are widely present in Lamiaceae, Asteraceae, Apiaceae, Rutaceae, Myricaceae, Dipterocarpaceae, Myristicaceae, Verbenaceae, Burseraceae, Poaceae and conifers. Their presence can be detected by smelling crushed leaves (or other plant parts).

Essential oils with monoterpenes are commonly used in aromatherapy and in phytomedicine to treat rheumatism, infections (bacterial, fungal), colds, restlessness, flatulence, intestinal spasms, as stomachicum and to improve taste. Essential oils are ingredients of many perfumes and of some natural insect repellents. Applied to the skin, monoterpenes and aliphatic hydrocarbons can cause hyperaemia; higher doses cause narcotic effects.

Monoterpenoids can be linear, monocyclic or bicyclic.

Examples of linear monoterpenoids.

Examples of monocyclic monoterpenoids.

Examples of bicyclic monoterpenes.

Myrtle (*Myrtus communis*): essential oil – 1,8 cineole

White peony (*Paeonia lactiflora*): roots – paeoniflorin

Citral is a mixture of two isomers, **citral A** (geranial) and **citral B** (neral). The essential oil of one chemotype of the Australian lemon myrtle tree (*Backhousia citriodora*) is almost pure citral. Citral is partly responsible (with **limonene**) for the strong lemon flavour of lemon verbena (*Aloysia citrodora*). Carvone occurs in nature as two isomers: (+)-carvone smells like caraway (it is the main compound in caraway seed oil from *Carum carvi*), while (−)-carvone has a spearmint smell and is the main constituent (50–80%) in spearmint oil (*Mentha spicata*).

Menthol accumulates in cornmint (*Mentha arvensis*) and peppermint (*M. ×piperita*). It induces a pronounced cooling sensation by triggering cold-sensitive receptors in the skin and mucosa. It also has analgesic effects by selectively activating κ-opioid receptors. **Piperitone** is used as a starting material for the production of synthetic menthol. It is obtained, in very high yield, from the essential oil of the broad-leaved peppermint gum (*Eucalyptus dives*).

Two of the most ubiquitous monoterpenoids are **camphor** and **1,8-cineole** (= eucalyptol). They often co-occur in essential oils and have been shown to exert a powerful synergistic antibacterial activity. The common bluegum (*Eucalyptus globulus*) is a well-known source of 1,8-cineole, while camphor is distilled from the wood of the camphor tree (*Cinnamomum camphora*).

Both enantiomers of **α-pinene** occur in nature; (−)-α-pinene is found in European pine trees. The compound is a bronchodilator and has anti-inflammatory and broad-spectrum antibiotic activities.

Thujone (in *Artemisia absinthium*, *Thuja* species, *Tanacetum vulgare*) contains a cyclopropane ring, which makes the molecule highly reactive. It appears that thujone can alkylate important proteins of the neuronal signal transduction, therefore causing neuronal disorder.

Sabinene (see *Juniperus sabina*) and **sabinol** are reactive monoterpenes with a highly reactive cyclopropane ring and/or with exocyclic or terminal methylene groups, as in **camphene**, pinocarvone or in **linalool**, which can bind to SH-groups of proteins and thus change their conformation.

Monoterpenes with a peroxide bridge, such as **ascaridol**, are reactive compounds, which can alkylate proteins.

Compounds with phenolic hydroxyl groups (such as **thymol** and **carvacrol**) or with an aldehyde function (such as citral, **citronellal**) can bind to proteins and exhibit pronounced antiseptic properties; they are active against bacteria and fungi.

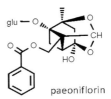
paeoniflorin

Monoterpene glycosides appear to be relatively rare but are found in several members of the Lamiaceae. Peony root (from the Chinese *P. lactiflora*) contains **paeoniflorin**, a monoterpenoid glucoside with antispasmodic, sedative and anti-inflammatory activities. It has been used as a dietary supplement.

Iridoids

Iridoids are related to monoterpenoids but they usually occur in plants as non-volatile glucosides. They were named for a genus of ants (*Iridomyrmex*) because they were first isolated from these ants, where they serve as defence compounds. Iridoids typically have a cyclopentane ring fused to a six-membered oxygen heterocycle. The biosynthesis proceeds from 10-oxogeranial, through reduction and cyclisation by the enzyme iridoid synthase. The two most common iridoids in plants are **aucubin** and **catalpol**. Iridoids often have a bitter taste and show a wide range of pharmaceutical activities, including analgesic, antihepatotoxic, anti-inflammatory, antimutagenic, antispasmodic, antitumour, antiviral, cardiovascular, choleretic, hypoglycaemic, immunomodulatory and laxative activities.

aucubin   catalpol

Secoiridoids are formed when a bond in the cyclopentane ring is cleaved. Secologanin derivatives (involved in ipecac alkaloid synthesis) are also formed in this way. More than 200 structures are known, distributed in the families Apocynaceae, Gentianaceae, Lamiaceae, Loganiaceae, Menyanthaceae, Plantaginaceae, Rubiaceae, Scrophulariaceae, Valerianaceae and Verbenaceae. Some of them, such as the gentiopicrosides, present in Gentianaceae and Menyanthaceae, exhibit an extremely bitter taste; they are used to improve digestion and to raise appetite in patients.

harpagoside

Iridoid glucosides, such as aucubin and **harpagoside**, are hydrolysed by β-glucosidase into an unstable aglycone. The lactol ring can open and produce a functional dialdehyde. Catalpol has a reactive epoxide ring in addition.

Several medicinal plants rich in iridoid glucosides have been used to treat infections, rheumatism and inflammations (*Plantago* species, *Harpagophytum procumbens*, *Scrophularia nodosa*, *Warburgia salutaris*).

amarogentin   gentiopicrin

Gentiana lutea is well known for its extremely bitter taste, resulting from the presence of **amarogentin** and **gentiopicrin**. These compounds are used as standard to determine bitter values.

The bitter taste of olives is due to **oleuropein**, a secoiridoid with demonstrated hypotensive activity. It occurs in all parts of the plant, including the fruits and leaves.

oleuropein   valerenic acid

The secoiridoids in *Valeriana officinalis* contribute to the sedating properties of the medicinally used drug. **Valerenic acid** is the main compound.

Sesquiterpenes

Sesquiterpenes are derived from the coupling of three isoprene units (3 × C5). More than 100 skeleton types are known, which can be acyclic, monocyclic, bicyclic or tricyclic. They often co-occur with monoterpenoids as the major components of essential oils, but are typically less volatile, resulting in longer retention times when analysed by gas-liquid chromatography. Sesquiterpenes are therefore found in the same plant families as the monoterpenoids, e.g. Apiaceae, Asteraceae, Burseraceae, Cupressaceae, Dipterocarpaceae, Lamiaceae, Myricaceae, Myristicaceae, Pinaceae, Poaceae, Rutaceae and Verbenaceae.

farnesol   zingiberene   α-bisabolol

An example of an acyclic sesquiterpenoid is **farnesol**, present in many essential oils (e.g. Tolu

Gardenia (*Gardenia jasminoides*): gardenoside (iridoids)

Figwort (*Scrophularia nodosa*): harpagoside (iridoids)

Valerian (*Valeriana officinalis*): valerenic acid (iridoids)

Cotton (*Gossypium hirsutum*): gossypol (sesquiterpene)

balsam, lemongrass and citronella). It is used in perfumery and to flavour cigarettes. **Zingiberene**, a component of ginger oil, is an example of a monocyclic sesquiterpenoid.

The dialdehydes **polygodial** and **warburganal** belong to the "drimane" group of sesquiterpenoids. They have a peppery taste and are recognised as the active principle in *Polygonum hydropiper*, *Drimys aromatica* and *Warburgia salutaris*. The dialdehyde can bind to proteins and form Schiff's bases with free amino groups, which appears to be the base for their pharmacological properties.

Bicyclic compounds also include the well-known **α-cadinene** (in oil of the cade juniper, *Juniperus oxycedrus*), **β-selinene** (celery seed oil), **(−)-β-caryophyllene** (clove oil) and **carotol** (car-

rot seed oil). Some sesquiterpenoids are highly aromatic, such as guaiazulene, a main compound in the bright blue oils of guaiac (see *Guaiacum officinale*) and the anti-inflammatory chamazulene from chamomile (*Matricaria chamomilla*) and wormwood (*Artemisia absinthium*). Chamomile has the monocyclic **α-(−)-bisabolol** as main compound in the oil (shown on page 52).

A wide diversity of tricyclic types occurs in nature. An example is the alcohol **patchoulol**, responsible for the typical scent of patchouli oil (*Pogostemon cablin*).

Sesquiterpene lactones

Sesquiterpene lactones are also formed from three isoprene units but they characteristically have a lactone ring. It is an exceptionally common and diverse group (more than 3 000 structures are known) that occurs mostly in Asteraceae and a few other families (Apiaceae, Magnoliaceae, Menispermaceae, Lauraceae) and ferns. The bitterness of many herbs can be ascribed to lactones.

These sesquiterpene lactones can bind to SH-groups of proteins via one or two exocyclic methylene groups and the enone configuration in the furan ring and are therefore pharmacologically active, often as anti-inflammatory agents. Some carry additional 1 or 2 epoxide functions, which make them even more reactive. Alkylated proteins can change their conformation and are no longer able to properly interact with substrates, ligands or other proteins. Also DNA can be alkylated, leading to mutations. Sesquiterpene lactones also bind glutathion (via SH-groups) and can deplete its content in the liver. As a consequence, these sesquiterpene lactones exhibit broad biological activities, including cytotoxic, antibiotic, anthelminthic, anti-inflammatory, phytotoxic, insecticidal and antifungal properties.

Several plants with sesquiterpene lactones have been used in traditional medicine or phytotherapy (*Achillea*, *Arnica*, *Matricaria*, *Parthenium*) because they have anti-inflammatory, expectorant, antibacterial, antifungal and antiparasitic properties. Many are structurally related to **costunolide** (first isolated from *Saussurea costus*, hence the name). It is a prototypical germacranolide.

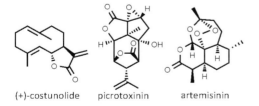

(+)-costunolide   picrotoxinin   artemisinin

The **picrotoxinin**-producing *Anamirta cocculus* has been used to treat vertigo.

Artemisinin from *Artemisia annua* has a reactive peroxide bridge. It has recently been developed into a potent antimalaria drug (artesunate), which is active against the dangerous *Plasmodium falciparum* that causes cerebral malaria. This compound is not part of the classical "sesquiterpene lactone" group.

Several examples of sesquiterpene lactones are presented here in alphabetical order.

anisatin   bilobalide

Anisatin is a lethal poison and insecticidal compound isolated from the shikimi plant (*Illicium anisatum*).

Bilobalide is a biologically active trilactone found in *Ginkgo biloba* leaves.

cnicin

Cnicin is the active compound in *Centaurea benedicta* (formerly known as *Cnicus benedictus*).

helenalin   alantolactone

Helenalin occurs in *Arnica montana* and *A. chamissonis* and is believed to be responsible for the anti-inflammatory, but also skin-irritant and toxic effects. Several other sesquiterpene lactones are also known for their toxic effects. Elfdock (*Inula helenium*) has **alantolactone** as main active compound.

lactucin   lactucopicrin

Lactucin and **lactucopicrin** occur in the milky latex of lettuce, chicory and dandelion. They are the main ingredients of lactucarium (lettuce opium), the dried latex that was once used as an official sedative and mild laxative in the USA and Britain. It was made from the latex of opium lettuce (*Lactuca virosa*) and included in throat lozenges and cough syrups. It was later found to be ineffective and lost its popularity. Since the hippie move-

Red stinkwood (*Prunus africana*): bark – β-sitosterol

Rapeseed [*Brassica rapa* (= *B. campestris*)]: seeds – campesterol

Pleurisy root (*Asclepias tuberosa*): cardiac glycosides

Red bryony (*Bryonia dioica*): roots – cucurbitacins

The best-known example of a dietary phytosterol is **β-sitosterol**, which occurs in pumpkin seeds (*Cucurbita pepo*) and many other plants (*Hypoxis hemerocallidea*, *Nigella sativa*, *Prunus africana*, *Serenoa repens*, *Urtica dioica*). These plants are traditionally used to treat benign prostate hyperplasia because the phytosterols are believed to inhibit the activity of the enzyme 5α-reductase and/or to decrease binding of dihydrotestosterone in the prostate. Another common phytosterol is **campesterol**, first found in rape seeds (*Brassica campestris*), hence the name. High levels (up to 100 mg/100 g) are found in canola oil and corn oil. It is thought that sitosterol and campesterol compete with cholesterol and therefore reduce the absorption of cholesterol in the human intestine.

Stigmasterol is used as a starting material in the semi-synthesis of cholesterol and cortisone. The latter has anti-inflammatory and painkilling activities and is used to treat many ailments. It is speculated that some plant steroids may mimic the effects of cortisone.

Among steroidal glycosides, the cucurbitacins (occurring in members of the Cucurbitaceae and a few other families) express substantial cytotoxic activities; they inhibit tumour growth *in vitro* and *in vivo*. Cucurbitacins have been used to treat nasopharyngeal carcinoma. They are highly cytotoxic as some of them block mitosis in metaphase by inhibiting microtubule formation.

A typical example is **cucurbitacin B**, present in *Citrullus colocynthis* and other members of the Cucurbitaceae. Drugs with cucurbitacins have been used to treat malaria, as emetic or anaesthetic (now obsolete), and in traditional medicine as diuretic, abortifacient and importantly as drastic laxative. Cucurbitacins irritate intestinal mucosa and cause release of water into the gut lumen. This in turn activates gut peristalsis and promotes diarrhoea. For topical use, *Bryonia* cucurbitacins have been applied to treat rheumatism and muscle pain.

Saponins

Saponins are the glycosides of triterpenes or steroids and include the group of cardiac glycosides and steroidal alkaloids. They are amphipathic glycosides, meaning that they have both hydrophilic (polar, "water-loving") and lipophilic (non-polar, "fat-loving") properties. The sugar part(s) of the saponin is hydrophilic, while the terpene part is lipophilic. Saponins are easily detected by the soap-like foaming that occurs when they are shaken in water. This has led to the so-called foam test (froth test), positive for saponins when the froth exceeds 20 mm in height and persists for 10 minutes or more. Monodesmosidic saponins have a single sugar chain, usually at C-3, while bidesmosidic saponins have two sugar chains, at C-3 and C-28.

oleanoglycotoxin A

An example of a monodesmosidic saponin with the sugar chain at C-3 is **oleanoglycotoxin A**, the active compound in endod (*Phytolacca dodecandra*).

asiaticoside

Asiaticoside (the main active compound in *Centella asiatica*) is also a monodesmosidic saponin but has the sugar chain attached at C-28.

Araloside, the main saponin of the Japanese angelica tree (*Aralia japonica*) is an example of a bidesmosidic saponin with sugar chains at both C-3 and C-28. It has anti-ulcer activity.

araloside A

When the sugars are removed (e.g. through acid hydrolysis), the aglycones are known as sapogenins.

Saponins are found in many plants but their name is derived from the well-known soapwort plant (*Saponaria officinalis*), the roots of which were traditionally used as a soap substitute and detergent for washing clothes. Another practical use of saponins is in the production of Turkish halva, where they are responsible for the unique texture that is so distinctive of these delicious sweets. Turkish soaproot and soaproot extracts are obtained from *Gypsophila graminifolia*, *G. bicolor*, *G. arrostii* and the closely related *Ankyropetalum gypsophiloides*.

The compounds are highly toxic for fish because they inhibit their respiration; therefore they have been traditionally used as fishing poisons. Saponins also kill water snails and have been employed to eliminate snails in tropical waters that transmit human parasites, such as *Schistosoma* (causing schistosomiasis, also known as bilharzia). An example of an African plant that is traditionally used as a soap substitute and as fish and snail poison in Ethiopia is endod (*Phytolacca dodecandra*).

Steroidal saponins are typical for several families of monocots, and are less frequent in dicots (Fabaceae, Scrophulariaceae, Plantaginaceae, Solanaceae, Araliaceae). Triterpene saponins are abundant in several dicot families, such as Caryophyllaceae, Ranunculaceae, Phytolaccaceae, Amaranthaceae, Primulaceae, Poaceae and Sapotaceae. They are absent in gymnosperms.

Some saponins are stored as bidesmosidic compounds in the vacuole, which are cleaved to the active monodesmosidic compounds by a β-glucosidase or an esterase upon wounding-induced decompartmentation. Monodesmosidic saponins are amphiphilic compounds, which can complex cholesterol in biomembranes with

Endod (*Phytolacca dodecandra*): saponins (soap substitute)

Elephant's foot (*Dioscorea elephantipes*): tuber – steroidal saponins

their lipophilic terpenoid moiety and bind to surface glycoproteins and glycolipids with their sugar side chain. This leads to a severe tension of the biomembrane and leakage. This activity can easily be demonstrated with erythrocytes, which lose their haemoglobin (haemolysis) when in contact with monodesmosidic saponins. This membrane activity is rather unspecific and affects a wide set of organisms, from microbes to animals. Therefore, saponins have been used in traditional medicine as anti-infecting agents.

Because saponins irritate the *nervus vagus* in the stomach, which induces the secretion of water in the bronchia, saponin-containing drugs are widely employed as secretolytic agents in phytomedicine (see *Hedera helix* and *Primula veris*, for example).

glycyrrhizin

In some cases, steroids, triterpenes and saponins structurally resemble endogenous anti-inflammatory hormones, e.g., glucocorticoids. The anti-inflammatory effects known from many medicinal plants could be due to a corticomimetic effect. A pronounced anti-inflammatory activity has been reported for **glycyrrhizin** (also known as glycyrrhizic acid) from liquorice (*Glycyrrhiza glabra*). This triterpene saponin is the main sweet-tasting substance in the rhizomes.

Steroidal saponins are important for the synthesis of steroid hormones that are used in the oral contraceptive pill, commonly known as "the pill". It has become a popular form of birth control in almost all countries since 1961 (except Japan) and is used by more than 100 million women worldwide.

diosgenin

Diosgenin, obtained by hydrolysis of dioscin (the main saponin in *Dioscorea villosa* and other species) is an example of a starting material for the commercial synthesis of steroids such as cortisone, pregnenolone and progesterone. Progesterone, for example, was used in early combined oral contraceptive pills. These steroids are nowadays semi-synthesised from phytosterols extracted from common sources such as soybeans.

Cardiac glycosides

Some saponins have additional functional groups, such as the **cardiac glycosides**. They can either be cardenolides (when they carry a five-membered cardenolide ring) or bufadienolides (with a six-membered bufadienolide ring). Cardenolides have been found in Plantaginaceae (formerly Scrophulariaceae; *Digitalis*), Apocynaceae (*Apocynum, Nerium, Strophanthus, Thevetia, Periploca, Xysmalobium*), Brassicaceae (*Erysimum, Cheiranthus*), Celastraceae (*Euonymus*), Convallariaceae (*Convallaria*) and Ranunculaceae (*Adonis*). Bufadienolides occur in Crassulaceae (*Cotyledon, Kalanchoe*), Hyacinthaceae (*Drimia*, formerly *Urginea*) and Ranunculaceae (*Helleborus*). Examples of cardenolides are shown below.

digitoxin

Digitoxin is the main poisonous cardenolide in the common purple foxglove (*Digitalis purpurea*).

oleandrin

Oleandrin is the main poisonous cardenolide in oleander (*Nerium oleander*).

uzarin

Uzarin is the active (non-toxic) cardenolide in the roots of uzara (*Xysmalobium undulatum*).

Two examples of bufadienolides are given below to show the typical six-membered bufadienolide ring in these compounds.

scillaren A

Scillaren A is the main toxic and active bufadienolide in the bulbs of the European squill (*Drimia maritima*).

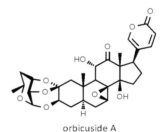

orbicuside A

Orbicuside A is a major poisonous bufadienolide from pig's ears or *plakkie* (*Cotyledon orbiculata*), a popular garden succulent. The compound causes a condition known as *krimpsiekte* in stock animals.

Although structurally different, all cardiac glycosides inhibit one of the most important molecular targets of animal cells, the Na^+, K^+-ATPase which builds up Na^+ and K^+ gradients, which are essential for transport activities of cells and neuronal signalling. Therefore, cardiac glycosides are strong neurotoxins, which cause death through cardiac and respiratory arrest. Cardiac glycosides are used in medicine to treat patients with cardiac insufficiency. They slow down the heartbeat and exhibit positive inotropic, positive bathmotropic, weakly negative chronotropic and dromotropic heart activity. Isolated cardiac glycosides are still used to treat patients with cardiac insufficiency; in phytomedicine standardised extracts of cardiac glycoside-producing plants are employed.

Uzara (*Xysmalobium undulatum*): roots – uzarin

Pig's ears (*Cotyledon orbiculata*): leaves – bufadienolides

Tetraterpenes

Carotenoids are formed from eight isoprene units (40 carbons) and represent the most important members of the tetraterpene group (more than 600 are known). They are organic pigments found in the chloroplasts and chromoplasts of plants. Carotenoids are highly lipophilic compounds and are always associated with biomembranes. In chloroplasts they serve as accessory pigments important for photosynthesis. They also protect against UV light. Carotenoids in food and medicinal drugs are employed as powerful antioxidants. Carotenoids (mainly β-carotene, α-carotene and cryptoxanthin) are the precursors for vitamin A in animals. Vitamin A is a group of unsaturated compounds (retinal, retinol and retinoic acid) along with several provitamin A carotenoids, among which β-carotene is the most important.

Carotenoids are divided into two groups: those with oxygen (**xanthophylls**) and those without (**carotenes**).

β-carotene (carotene)

α-carotene

lycopene

The main dietary carotenes are **β-carotene** (often referred to simply as carotene, because it is the major carotenoid), **α-carotene** and **lycopene**. Carrot (*Daucus carota*) is a major source of dietary carotene. Lycopene is the major red pigment in tomatoes (*Lycopersicon esculentum* or *Solanum lycopersicum*).

lutein

zeaxanthin

cryptoxanthin

crocetin

Examples of xanthophylls include **lutein**, **zeaxanthin** and **cryptoxanthin**. The flavour of saffron (*Crocus sativus*) is due to crocin, a digentiobioside ester of the carotenoid **crocetin**.

all-trans-retinal   retinol

all-trans-retinoic acid

The carotenoids are used to produce **retinal** (a light sensor in the rhodopsin complex) and **retinoic acid** (retinoids bind to nuclear receptors and are local mediators of vertebrate development).

Polyterpenes

Polyterpenes, consisting of 100 to 10 000 isoprene units, are prominent in latex of Euphorbiaceae, Moraceae, Apocynaceae, Sapotaceae and Asteraceae. Some polyterpenes are used commercially, such as rubber (from *Hevea brasiliensis*, Euphorbiaceae) or gutta-percha.

Phenolics

A large group of secondary metabolites, which are produced by most plants, are phenolics. They contain one or several aromatic rings carrying phenolic hydroxy groups. Phenolic substances often occur as glycosides and tend to be water soluble. They are all aromatic and therefore show intense absorption of light in the UV region of the spectrum. This is a useful means of detecting, identifying and quantifying phenolic compounds. The group includes simple phenols, phenylpropanoids, flavonoids and phenolic quinones, as well as polymeric materials such as tannins and lignins.

The phenolic hydroxy groups of these compounds can dissociate in negatively charged phenolate ions under physiological conditions. Phenolic hydroxy groups can thus form both hydrogen and ionic bonds with many proteins and peptides involved in health disorders. The higher the number of hydroxy groups, the stronger the astringent and denaturing effect. A common chemical property is the scavenging of oxygen radicals and therefore many phenolics exert antioxidant activity. Polyphenols (with several phenolic rings) are present in most drugs used in phytotherapy and apparently are responsible for a wide array of pharmacological properties, including antioxidant, anti-inflammatory, sedating, wound-healing, antimicrobial and antiviral activities.

Phenylpropanoids

Simple phenols with pronounced antioxidant activity include **phenol** itself, *p*-**cresol**, **catechol**, **gallic acid**, **guaiacol**, **orcinol**, **4-methylcatechol**, **phloroglucinol**, **pyrogallol**, **syringol** (2,6-dimethoxyphenol) and **resorcinol**. They carry two or three phenolic hydroxy groups, which are either free or methylated.

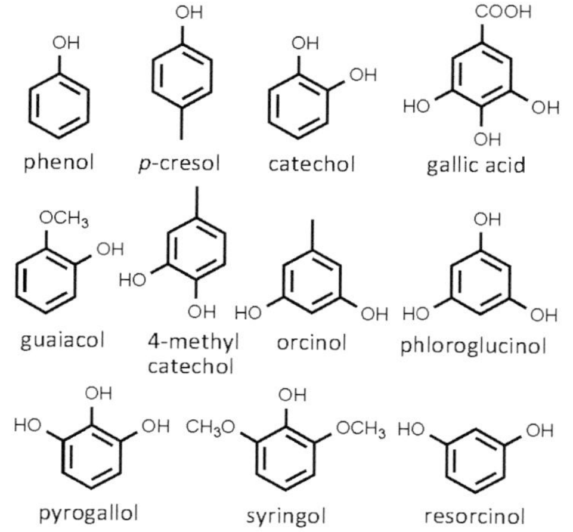

Medicinally important phenylpropanoids with a shortened side chain include **salicylic acid**, saligenin and the corresponding glucoside **salicin**.

Because they inhibit a key enzyme of prostaglandin biosynthesis, i.e. cyclooxygenase, they have been used in the treatment of inflammation, fever and chronic pain. These compounds are known from willows (*Salix purpurea*), *Populus*, *Filipendula ulmaria*, *Primula veris* and *Viola tricolor*.

Some lipophilic and aromatic **phenylpropanoids** include **myristicin**, **safrole**, **eugenol**, **apiole**, **ß-asarone**, **elemicin** and **estragole**, which can be found in essential oils.

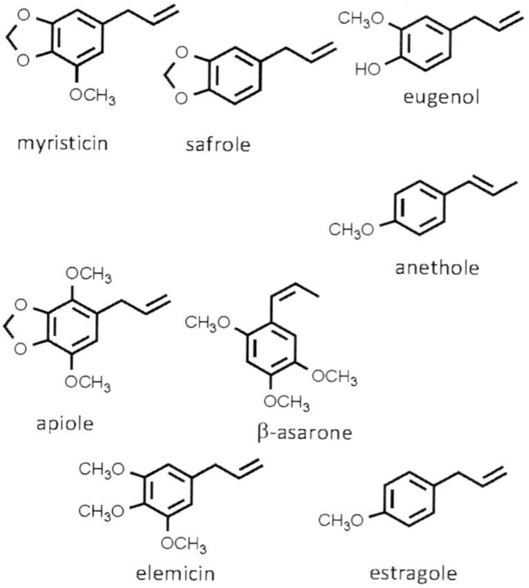

Phenylpropanoids with a terminal methylene group as shown here can react with SH-groups of proteins. In the liver, these compounds are converted to epoxides, which can alkylate proteins and

Sassafras (*Sassafras albidum*): leaves – safrole

Sweet woodruff (*Galium odoratum*): leaves – coumarin

DNA. Therefore, they are potentially mutagenic and tumours have been observed in animal experiments. In particular, myristicin inhibits MAO, which induces an increase of biogenic amine neurotransmitters, such as dopamine, serotonin and noradrenaline. Psychotropic effects resemble those of amphetamine. Eugenol is antiseptic and analgesic and has been widely used in dentistry. Another well-known phenylpropanoid is **anethole** (see *Pimpinella anisum*).

quinic acid

chlorogenic acid

neochlorogenic acid

cichoric acid

Chlorogenic acid and related **quinic acid** esters of caffeic acid occur in some medicinal plants that are used as general tonics, including *Centella asiatica* and *Echinacea* species. **Neochlorogenic acid** occurs in dried fruits and is believed to be responsible for their mild laxative effect. The structurally related **cichoric acid** occurs in *Cichorium intybus* and in *Echinacea purpurea* (but not any of the other *Echinacea* species). Phenolic acids are treated in more detail in the section on organic acids.

rosmarinic acid

coumaroylputrescine

Phenylpropanoids can also be conjugated with a second phenylpropanoid, such as in **rosmarinic acid** or with amines, such as **coumaroylputrescine**. Rosmarinic acid (common in Lamiaceae) bears a number of phenolic hydroxy groups with tannin-like activity (anti-inflammatory, antiviral).

urushiol I

Some phenols carry long alkyl and alkenyl side chains. Alkyl and alkenyl phenols such as urushiol are abundant in Anacardiaceae, Hydrophyllaceae, Proteaceae, *Ginkgo* and *Philodendron*. The example shown here is **urushiol I**, found as one of a mixture of urushiols in poison ivy (*Rhus toxicodendron*). Alkyl phenols are extremely allergenic compounds that are responsible for over a million poisoning cases (*Rhus* dermatitis) reported in the USA. Contact with the eye is extremely hazardous and can lead to blindness.

Phenylpropanoids occur in abundance in the ginger family (Zingiberaceae). The pungent taste (and medicinal value) of ginger (*Zingiber officinale*) is due to gingerols (such as **6-gingerol**) in the intact rhizome and derivatives (shogaols and **zingerone**) that form when ginger is dried or cooked.

Diarylheptanoids occur in large numbers in ginger and related plants. These compounds comprise two aromatic rings connected by a chain of seven carbons (heptane). An example is **gingerenone A**, a major compound in ginger. The related turmeric (*Curcuma longa*) has curcumin, a major food colouring with demonstrated medicinal activities.

Coumarins and furanocoumarins

Phenylpropanoids serve as building blocks for coumarins and furanocoumarins of which over 700 structures have been determined. These compounds are named after *coumarou*, the French name for the tonka bean (*Dipteryx odorata*) from which coumarin was first isolated. Coumarins can reach concentrations of up to 2% in plants and are common in the Apiaceae (most genera), Fabaceae (e.g., *Dipteryx odorata*, *Melilotus officinalis*), Poaceae (e.g., *Anthoxanthum odoratum*), Rubiaceae (e.g., *Galium odoratum*). In phytomedicine they are used because of anti-inflammatory and antimicrobial properties (*Melilotus*). Coumarins are aromatic and therefore used in cosmetics and in beverages. They are often components of essential oils.

Simple coumarins include **coumarin** itself, present in tonka bean, sweet clover (*Melilotus officinalis*); **umbelliferone**, in many members of the Apiaceae (Umbelliferae) such as angelica, coriander and carrot, but also in mouse-ear hawkweed (*Hieracium pilosella*, Asteraceae) and hydrangea (*Hydrangea macrophylla*, Hydrangeaceae); **umckalin**, present in *Pelargonium sidoides* (the additional presence of coumarin sulfates distinguishes this species from *P. reniforme*). These coumarins also occur as glycosides, e.g. **aesculin** in horse chestnut (*Aesculus hippocastanum*).

Furanocoumarins usually have a third furane ring that derives from active isoprene. Linear (psoralen-type) or angular (angelicin-type) furanocoumarins are distinguished.

Linear furanocoumarins include the widely distributed **psoralen**, **bergapten** and **xanthotoxin**, as well as **khellin** (in *Visnaga daucoides*).

Examples of the angular type include **angelicin** (from *Angelica archangelica*) and **visnadin** (from *Visnaga daucoides*).

The furanocoumarins occur in aerial parts such as leaves and fruits but also in roots and rhizomes. They are abundant in Apiaceae (contents up to 4%), but also present in certain genera of the Fabaceae (e.g., *Psoralea bituminosa*) and Rutaceae. The lipophilic and planar furanocoumarins can intercalate DNA and upon illumination with UV light can form cross-links with DNA bases, but also with proteins. They are therefore mutagenic and possibly carcinogenic. In medicine, furanocoumarins (such as 8-methoxypsoralen, 8-MOP) are employed for the treatment of psoriasis and vitiligo because they can kill proliferating keratocytes in the skin upon UV exposure. This treatment brings some relief for psoriasis patients.

Lignans and lignin

Phenylpropanoids can form complex dimeric structures, so-called lignans.

podophyllotoxin

Podophyllotoxin, which occurs in members of the genera *Podophyllum* (Berberidaceae), *Linum* (Linaceae) and *Anthriscus* (Apiaceae), is a potent inhibitor of microtubule formation and thus prevents cell division.

pinoresinol

Pinoresinol and related compounds are inhibitors of cAMP phosphodiesterase, cytotoxic, insecticidal and immune modulating.

silybin

The lignans from *Silybum marianum* (**silybinin**, silandrin, silychristin) have antihepatotoxic properties and the product is used to treat *Amanita* poisoning and liver cirrhosis.

Condensation of phenylpropanoids generates the complex lignin macromolecules that are important for the mechanical stability of plants, but also show some antimicrobial effects.

Flavonoids and anthocyanins

Phenylpropanoids can condense with a polyketide moiety to form flavonoids, stilbenes, chalcones, catechins and anthocyanins. These compounds are characterised by two aromatic rings that carry several phenolic hydroxy or methoxy groups. In addition, they often occur as glycosides and are stored in vacuoles. Flavonoids are active ingredients of many phytopharmaceuticals.

Flavonoids have a typical **C15 nucleus** with three rings. The flavonoid nucleus is usually attached to a sugar and the compounds occur mostly as glycosides in the vacuoles of cells. The main classes of flavonoids are flavones, flavanones, flavonols, chalcones and aurones, and isoflavones and isoflavonoids.

The flavonoid nucleus

apigenin (a flavone)

Flavones are the most basic type of flavonoid and occur in many angiosperms. **Apigenin** and luteolin are well-known examples.

naringenin (a flavanone)

hesperitin (a flavanone)

Flavanones are similar to flavones but differ in the absence of a double bond in the 2,3-position. They are common in many plant families (Asteraceae, Fabaceae, Rosaceae and Rutaceae). **Naringenin** is an example – it is the dominant flavanone in grapefruits (an aglycone of naringin). Another is **hesperitin**, the aglycone of hesperidin. It is released through acid hydrolysis when citrus fruits are ingested.

Isoflavones and isoflavonoids are typical secondary metabolites of the legumes (subfamily Papilionoideae). They resemble the female sex hormone estradiol. Isoflavones can exhibit oestrogenic properties and inhibit tyrosine kinases. Because of these properties they are often regarded as useful compounds that might play a role in the prevention of certain cancers, and for women with menopause or osteoporosis problems.

genistein (an isoflavone)

daidzein (an isoflavone)

rotenone (an isoflavanoid)

Isoflavones from soy bean (*Glycine max*) and red clover (*Trifolium pratense*) are marketed as nutraceuticals. The two main compounds in these preparations are **genistein** and **daidzein**. **Rotenone** is an isoflavonoid that inhibits the mitochondrial respiratory chain and is therefore highly toxic and used as an insecticide (traditionally as a fish poison).

kaempferol (a flavonol)

quercetin (a flavonol)

Flavonols are very common in plants. They are characterised by a hydroxy group in the 3-position. Glycosides of **kaempferol** and **quercetin** are frequently found in many medicinal plants.

resveratrol (a stilbenoid)

Stilbenoids are hydroxylated derivatives of stilbene with a C6-C2-C6 structure that are biogenetically related to the chalcones. Stilbenes such as **resveratrol** (present in red wine) have antioxidant, antibacterial and antifungal activities, and are present in several drugs and nutraceuticals.

Chalcones are characterised by an open C3 heterocyclic ring.

phlorizin (a chalcone)

aspalathin (a chalcone)

isoliquiritigenin (a chalcone)

Examples include **aspalathin**, the main phenolic compound in rooibos tea (*Aspalathus linearis*), the *O*-glycoside **phlorizin** from *Acorus*, *Pieris* and *Rhododendron* (it inhibits glucose transport at biomembranes) and **isoliquiritigenin** (it inhibits mitochondrial monoamine oxidase and uncouples mitochondrial oxidative phosphorylation). Glyceollin II, a prenylated pterocarpan and phytoalexin with anti-oestrogenic activity, has the same activity.

taxifolin-3-O-acetate (a dihydroflavonol)

6-methoxytaxifoline (a dihydroflavonol)

Dihydroflavonols such as **taxifolin 3-*O*-acetate**, **6-methoxytaxifolin** and 6-methoxyaromadendrin 3-*O*-acetate have a sweet taste. This is unlike many flavanones such as naringin, neoeriocitrin and neohesperidin that are typically very bitter.

Anthocyanins are the dominant flower and leaf pigments in plants that give red, pink, blue and purple colours, including autumn colours. The only exception is certain plant families of the order Caryophyllales, where the bright colours are due to betacyanins or betalains (indole-derived pigments) that contain nitrogen.

The bright red colour of beetroot, for example, is due to **betanin** (beetroot red). It is used as a food additive. The aglycone is called betanidin.

There are six very common anthocyanin aglycones (called anthocyanidins) in nature, three with hydroxy groups on the B-ring and three with methoxy groups. These are **cyanidin, pelargonidin** and **delphinidin** (hydroxylated) or **peonidin, petunidin** and **malvidin** (methoxylated).

The colour of anthocyanins depends on the degree of glycosylation, hydrogen ion concentration and the presence of certain metals [e.g., aluminium (aluminum) ions] in the vacuole. Parallel to a change in pH of the vacuole in developing flowers, a colour change from pink to dark blue can be observed in several species of the Boraginaceae (e.g., *Symphytum, Echium*). Anthocyanins are active antioxidants and are therefore used in phytomedicine or nutraceuticals to prevent ROS-related health disorders (anthocyanin-rich fruits and fruit juices from *Aronia, Vaccinium, Punica, Vitis* and others). ROS refers to reactive oxygen species that cause damage to cells and DNA.

Catechins and tannins

Catechins form a special class of flavonoids, which often dimerise or even polymerise to form procyanidins and oligomeric procyanidins.

The basic structure is a **flavan-3-ol**. **Procyanidin B-2** is an example of a dimeric catechin found in *Crataegus* flowers and in grape leaves.

Common catechins include **catechin** and **epicatechin**, both found in tea and cacao. The conjugates (which cannot be hydrolysed; "non-hydrolysable or condensed tannins") are characterised by a large number of hydroxy groups. The phenolic hydroxy groups can interact with proteins to form hydrogen and ionic bonds and possibly even covalent bonds. If more than 10 hydroxy groups are present these compounds act as "tannins". Non-hydrolysable tannins are also called condensed tannins (or proanthocyanidins, polyflavonoid tannins, catechol-type tannins, pyrocatecholic type tannins or flavolans). They are polymers formed by the condensation of flavans but they have no sugar residues. The term proanthocyanidin is appropriate because these polymers yield anthocyanidins when they are depolymerised. The different types are called procyanidins, propelargonidins, prodelphinidins, and so on, depending on the units. The tannin-protein interactions are a base for the utilisation of plants with catechins in phytotherapy (e.g., *Crataegus monogyna* in patients with heart problems).

gallic acid

Another important group of tannins is hydrolysable. They represent esters of **gallic acid** and sugars; in addition several moieties of gallic acid can be present that are also linked by ester bonds. These gallotannins are widely distributed in plants, often in bark, leaves and fruits. Gallotannins, which can additionally be condensed with catechins, contain a large number of phenolic hydroxy groups so that they can form stable protein-tannin complexes and thus interact with a wide variety of protein targets in microbes and animals.

pentagalloylglucose

An example of a hydrolysable tannin is **pentagalloylglucose**, found in pomegranate (*Punica granatum*) and several other plants.

Ellagitannins differ from gallotannins because their galloyl groups are linked through C-C bonds. In this case the acid component is **chebulic acid** or **ellagic acid**.

ellagic acid   chebulic acid

Examples include **punicalin** from pomegranate (*Punica granatum*) and **agrimoniin** from agrimony (*Agrimonia eupatoria*) and wild strawberry (*Fragaria vesca*).

punicalin

agrimoniin

Tannins are known for their tanning properties (i.e. to turn raw hide into leather by forming a resistant layer on the collagen fibres). They are also traditionally used as antidiarrhoeals and externally as vasoconstrictors. Tannins are strong antioxidants, with anti-inflammatory, antidiarrhoeal, cytotoxic, antiparasitic, antibacterial, antifungal and antiviral activities. Several medicinal plants (*Quercus*, *Krameria*, *Alchemilla*, *Agrimonia*, *Potentilla*) are used internally and externally to treat inflammation and infection.

Small reactive molecules

This short section accommodates some important small molecules that do not fit comfortably elsewhere. They have no direct relation to phenolics. Included here are ranunculin, tuliposide and ethanol (alcohol).

ranunculin   protoanemonin

Ranunculin is a characteristic secondary metabolite of Ranunculaceae. When plant tissue is damaged, ranunculin is converted to **protoanemonin**.

Orange lily (*Clivia miniata*): lycorine

Autumn crocus (*Colchicum autumnale*): colchicine

Australian chestnut (*Castanospermum australe*): seeds – castanospermine

Californian poppy (*Eschscholzia californica*): roots – isoquinoline alkaloids

Colchicine

Colchicine and related alkaloids are typical secondary metabolites of plants in the genera *Colchicum*, *Gloriosa* and a few other members of the family Colchicaceae (formerly associated with Liliaceae).

colchicine

The molecular target of colchicine is tubulin; it inhibits the polymerisation and depolymerisation of microtubules which are necessary for cell division and intracellular transport of vesicles. Colchicine inhibits the synthesis of collagen and activates collagenase. Colchicine has been used against fast-dividing cancer cells, but its toxicity prevents a general application. In modern medicine, colchicine is prescribed in cases of acute gout as it prevents macrophages from migrating to inflamed joints.

Diterpene alkaloids

Aconitine from *Aconitum* species and **protoveratrine B** from *Veratrum* species are potent activators of Na^+-channels that are essential for neuronal signalling. If these ion channels are completely activated, the action potential from nerves to muscles is no longer transmitted, leading to a complete arrest of cardiac and skeletal muscles.

aconitine

Aconitine and protoveratrine B first activate and then paralyse the sensible nerve endings and neuromuscular plates. Aconitine also exerts analgesic properties and has been used to treat neuronal pain, such as caused from irritation of the trigeminus nerve. Extracts from *Aconitum* have been widely used as arrow poison, deadly poison and in witch ointments for thousands of years in Europe and Asia.

protoveratrine B

Another diterpene alkaloid is **paclitaxel**, (Taxol®) which can be isolated from several yew species (including the North American *Taxus brevifolia* and the European *T. baccata*). Taxol® stabilises microtubules and thus blocks cell division in the late G2 phase; because of these properties, Taxol® has been used for almost 20 years with great success in the chemotherapy of various tumours.

Docetaxel (trade name Taxotere or Docecad) is a semi-synthetic analogue of paclitaxel, developed because of the scarcity of the latter. It is an esterified product of 10-deacetyl baccatin III, a starting material which can be extracted from the leaves of the common and readily available European yew (*Taxus baccata*).

paclitaxel (Taxol®)

dodetaxel

Ergot alkaloids

Included here are the clavine alkaloids (**agroclavine** and elymoclavine), lysergic acid amides (ergine, **ergometrine** and more complex peptide alkaloids, such as **ergotamine** and ergocristine).

agroclavine

ergometrine

ergotamine

LSD

Ergot alkaloids are produced by a symbiotic fungus *Claviceps purpurea*, and more than 40 other species which are symbionts on grasses (tribes Festucaceae, Hordeae, Avenae, Agrosteae). Rye is especially affected. Ergot alkaloids are also found in some Convolvulaceae (*Argyreia, Ipomoea, Rivea corymbosa, Stictocardia tiliafolia*) which carry the fungi as endophytes.

These alkaloids modulate the activity of noradrenaline, serotonin and dopamine receptors as agonists, partial agonists but also antagonists. Consequences are contraction of smooth muscles of peripheral blood vessels (causing gangrene), or permanent contraction of uterine muscles (causing abortion). By blocking alpha-adrenergic receptors, the alkaloids can induce spasmolysis (relaxation of smooth muscles). They inhibit serotonin receptors but stimulate dopamine receptors. Ergometrine (an α-receptor agonist) is used in obstetrics to stop bleeding after birth or abortion. Ergotamine (antagonist at noradrenaline and 5-HT receptor; agonist at dopamine receptor) is used to treat migraine. Ergocornine reduces the secretion of prolactin and inhibits nidation as well as lactation. **LSD** (*N,N*-diallyllysergic acid amide), which is a synthetic derivate of ergot alkaloids, is one of the strongest hallucinogens.

Poisoning with ergot alkaloid-contaminated cereals and flour causes the dramatic and cruel effects of ergotism which has been documented in many paintings of the Old Masters. The hallucinogenic Mexican drug "ololiuqui" is composed of ergot alkaloids of *Rivea corymbosa, Ipomoea argyrophylla, I. violacea* and other *Ipomoea* species.

Indole alkaloids

Indole alkaloids (including monoterpene indole alkaloids) occur mainly in four plant families – the Apocynaceae, Loganiaceae, Gelsemiaceae and Rubiaceae.

ajmaline

ajmalicine

Ajmaline from *Rauvolfia serpentina* blocks sodium channels and has therefore antiarrhythmic properties because it lowers cardiac excitability. It has negative inotropic properties and is used medicinally to treat tachycardial arrhythmia, extra systoles, fibrillation and *angina pectoris*. **Ajmalicine** (also from *Rauvolfia serpentina*) has a pronounced dilatatoric activity in blood vessels, which causes hypotension. Ajmalicine is used as a tranquilliser and as an antihypertensive to improve cerebral blood circulation.

C-toxiferine I

C-Toxiferine I and II from *Strychnos* are neuromuscular blocking agents, thus highly toxic and used as an arrow poison. They are strong inhibitors of nicotinic AChR at the neuromuscular plate and cause paralysis of muscle cells.

ibogaine

Ibogaine from *Tabernanthe iboga* is a CNS stimulant with anticonvulsant and hallucinogenic properties.

Physostigmine, eseridine and related compounds from *Physostigma venenosum* (calabar beans) are strong inhibitors of cholinesterase with wide-ranging parasympathetic activities.

physostigmine

Physostigmine is used as a miotic in eye treatments and in the therapy of Alzheimer's disease. It is highly toxic and calabar beans were used as an ordeal poison in West Africa.

harmine

Harman or β-carboline alkaloids occur, among others, in Malpighiaceae (*Banisteriopsis*), Zygophyllaceae (*Peganum*, *Zygophyllum*) and Rutaceae (*Clausena*, *Murraya*). β-Carboline alkaloids are inhibitors of MAO and agonists at serotonin receptors. Since they enhance serotonin activity, they exhibit substantial hallucinogenic activities and might be useful to treat patients with depression. An example is **harmine**, the main compound in *Peganum harmala*.

vinblastine

Dimeric *Vinca* alkaloids (vincristine, **vinblastine**, leurosine) from *Catharanthus roseus* inhibit tubulin polymerisation and intercalate DNA. As a consequence they effectively block cell division and are therefore important drugs used in cancer therapy.

camptothecin

Camptothecin, an inhibitor of DNA topoisomerase used in cancer therapy, is mainly produced from *Camptotheca acuminata* (but is also found in some genera of Icacinaceae, Rubiaceae, Apocynaceae and Gelsemiaceae).

reserpine

Reserpine and related alkaloids from *Rauvolfia serpentina* inhibit transporters for neurotransmitters at vesicle membranes and thus act as an antihypertensive and tranquilliser.

strychnine

Strychnine from *Strychnos nux-vomica* is an antagonist at the glycine-gated chloride channel. It is a CNS stimulant and extremely toxic.

mesembrine mesembrenone mesembrenol

Mesembrine, a simple indole alkaloid from *Sceletium tortuosum* (= *Mesembryanthemum tortuosum*), is a narcotic with cocaine-like activities and has been used as an antidepressant. Extracts rich in **mesembrenone** and **mesembrenol** enhance cognitive function.

gelsemine

Gelsemine and gelsemicine are CNS active and highly toxic. They are the main alkaloids of *Gelsemium sempervirens*.

Indolizidine alkaloids

Indolizidine is an isomer of indole. It forms the base of several alkaloids, mainly found in the legume family (Fabaceae). The compounds are also called polyhydroxy alkaloids or sugar-shaped alkaloids because they mimic sugar in structure (but with nitrogen in the place of oxygen). They also behave like sugars in solution and are therefore missed with the usual method of alkaloid extraction (cation exchange resin is therefore used).

castanospermine swainsonine

Castanospermine is the main alkaloid in seeds of the Australian chestnut tree, *Castanospermum australe*. It is an inhibitor of glucosidase enzymes and has been studied for its antiviral activity (also against the HIV virus). **Swainsonine** is the poisonous compound in locoweed, responsible for severe stock losses in North America (mainly the western parts of the USA). Locoweed refers to pasture plants that contain swainsonine, which include several species of the legume genera *Astragalus* and *Oxytropis* (and *Swainsonia* in Australia). Animals develop a condition called "locoism" in North America (and "pea struck" in Australia).

Isoquinoline alkaloids

Isoquinoline alkaloids include protoberberine, aporphine and morphinane alkaloids. Isoquinoline alkaloids are common in genera of the Papaveraceae, Annonaceae, Ranunculaceae, Berberidaceae, Monimiaceae, Menispermaceae, Lauraceae, Rutaceae and Magnoliaceae.

berberine sanguinarine

Many protoberberine and benzophenanthridine alkaloids interfere with neuroreceptors and DNA (several are strong intercalators). The intercalating alkaloids (e.g. **berberine**, **sanguinarine**) show pronounced antibacterial, antiviral and cytotoxic properties. Extracts of *Sanguinaria canadensis*, which are rich in the benzophenanthridine alkaloid sanguinarine, have been included in mouthwashes and toothpaste.

Chelidonium majus has been used in traditional medicine and phytomedicine as cholagogue, spasmolytic, diuretic and analgesic drug or to treat warts. **Chelidonine** has been employed as a painkiller to treat abdominal pain, and to treat spasms and asthma. Extracts of *Eschscholzia californica*, which are rich in aporphine, protoberberine and benzophenanthridine alkaloids, have been employed as a mild psychoactive drug to induce euphoria.

chelidonine

boldine

The aporphine **boldine** (from *Peumus boldus*) is used to treat hepatic dysfunction and cholelithiasis.

emetine

cephaeline

Emetine and **cephaeline** from (*Psychotria ipecacuanha*) are potent inhibitors of ribosomal protein synthesis; they have been used as emetics, expectorants and anti-amoebics.

cepharanthine

erysovine

Cepharanthine, a bisbenzylisoquinoline from *Stephania*, has been used to treat tuberculosis and leprosy. *Erythrina* alkaloids block signal transduction at the neuromuscular plate and have been used as curare substitute. An example is **erysovine**, a major alkaloid of *E. lysistemon* and *E. caffra*.

Tubocurarine and other bisbenzylisoquinolines from *Chondrodendron* and *Ocotea* have been used traditionally as arrow poison but also in surgery as muscle relaxant (inhibition of nAChR).

tubocurarine

These alkaloids have the advantage that the prey animal (monkey or parrot) relaxes its grip when it dies, so that it can be easily retrieved. Furthermore, the alkaloids are only poisonous when injected and are quite harmless when ingested.

papaverine

Papaverine (from several *Papaver* species) inhibits phosphodiesterase and thus acts as smooth muscle relaxant, vasodilator and spasmolytic.

morphine

codeine

Morphinane alkaloids are typical for members of *Papaver somniferum* and *P. bracteatum*. Morphine causes central analgesia, euphoria and sedation. **Morphine** is an agonist of endorphine receptors in the brain and other organs and promotes powerful sleep-inducing, analgesic and hallucinogenic effects. It is used in standardised modern medicines intended for oral and parenteral use – mainly to treat intense pain (e.g. in cancer patients). **Codeine** is an effective painkiller (though less active than morphine, but also less addictive); it sedates the cough centre and is widely used as antitussive agent. Morphine and other morphinane alkaloids show addictive properties.

Phenylpropylamines

This group of bioactive amines with pronounced pharmacological activity includes **cathinone** (from *Catha edulis*), **ephedrine** (from several *Ephedra* species) and **mescaline** (*Lophophora williamsii* and other cacti).

Cathinone and ephedrine structurally resemble **amphetamine** and act in a similar way as sympathomimetics. These alkaloids stimulate α- and β-adrenergic dopaminergic receptors by stimulating the release of noradrenaline and dopamine from catecholic synapses and inhibiting their re-uptake. Ephedrine causes vasoconstriction, hypertension, bronchial dilatation and heart stimulation. Plants with ephedrine or cathinone reduce hunger sensation and have been used as appetite depressant and stimulant. Ephedrine has been used medicinally to treat asthma, sinusitis and rhinitis. Mescaline is a psychomimetic; it is a CNS depressant and hallucinogenic in high doses.

Piperidine alkaloids

Arecoline and **arecaidine** from betel nut (*Areca catechu*) exhibit parasympathetic activities and act as a central stimulant. Betel is widely used in Southeast Asia as a masticatory.

Piperine is the pungent principle of *Piper nigrum* and other species. *Piper* fruits are widely used as hot spice and sometimes as insecticide.

Coniine is a famous toxin from *Conium maculatum* which acts as a muscarinergic agonist. It causes ascending paralysis, which starts at the extremities of the arms and legs and ends with respiratory failure and death. *Conium* alkaloids are extremely toxic and teratogenic in livestock.

Lobeline occurs in *Lobelia* species and has been used in the treatment of asthma and as anti-smoking drug.

Pelletierine from *Punica granatum* has been used against intestinal tapeworms.

Ammodendrine often co-occurs with quinolizidine alkaloids in members of the Fabaceae. It can cause malformation in cattle if pregnant cows feed on plants which contain this alkaloid.

Purine alkaloids

Caffeine, **theophylline** and **theobromine** are produced by *Coffea arabica*, *Cola acuminata*, *Cola nitida*, *Theobroma cacao*, *Paullinia cupana*, *Ilex paraguariensis* and *Camellia sinensis*.

The purine alkaloids function as central nervous system stimulants, conferring wakefulness and enhanced mental activity. Caffeine inhibits cAMP phosphodiesterase and adenosine receptors. As a consequence dopamine is released and many brain parts become activated. These alkaloids are cardiac stimulants, vasodilators and smooth muscle relaxants. Extracts with purine alkaloids are widely used by humans as stimulants; caffeine is incorporated into numerous formulations employed against fever, pain and flu symptoms.

Pyrrolidine alkaloids

Nicotine (from *Nicotiana tabacum*) is an agonist at nACh-receptors and functions as a CNS stimulant with addictive and tranquillising properties. Today it is also used in electronic cigarettes ("e-cigarettes"). Before the availability of synthetic insecticides, nicotine was widely used as a natural insecticide in agriculture.

nicotine

Pyrrolizidine alkaloids

Pyrrolizidine alkaloids are produced from nearly all members of the Boraginaceae, several Asteraceae (subfamily Senecioninae) and Fabaceae (tribe Crotalarieae).

Examples of well-known compounds include **senecionine** (produced by many *Senecio* species and also Fabaceae – *Crotalaria* and *Lotononis* species) and **heliotrine**, a typical product from Boraginaceae genera such as *Amsinckia* and *Heliotropium*.

heliotrine   senecionine

Pyrrolizidine alkaloids are activated in the liver of humans or animals to reactive pyrroles (dehydropyrrolizidines) that can alkylate DNA bases. These alkylations can lead to mutation and cell death (especially in the liver). Furthermore, mutations can lead to malformations in pregnant animals and humans, and to cancer of liver, kidneys and lungs.

Several pyrrolizidine alkaloid-containing plants are used in traditional phytomedicine to treat bleeding or diabetes or as general herbal teas (*Senecio, Petasites, Heliotropium, Crotalaria*); *Symphytum officinale* and other Boraginaceae are used to treat wounds, broken or injured bones. Others, such as comfrey (*Symphytum ×uplandicum*) are regularly supplied on local markets as "healthy" salad ingredients. Drugs containing pyrrolizidines are usually banned as medicines.

Quinolizidine alkaloids

Quinolizidine alkaloids such as **sparteine**, **lupanine**, **lupinine**, **cytisine** and **anagyrine** are common secondary metabolites in genistoid legumes (Fabaceae). They affect acetylcholine receptors and ion channels; they are poisonous neurotoxins for animals.

(−)-sparteine   (+)-sparteine (=pachycarpine)

(+)-lupanine   (−)-lupanine

lupinine   cytisine   anagyrine

Sparteine from *Cytisus scoparius* has been employed medicinally to treat heart arrhythmia (Na^+-channel blocker) and during childbirth (inducing uterus contraction). Plants with anagyrine can cause malformations ("crooked calf disease") if pregnant animals feed on plants (such as lupins) containing it.

Quinoline alkaloids

Quinoline alkaloids are here considered to include acridone alkaloids. Medically important quinolone alkaloids occur in Rutaceae, Acanthaceae and Rubiaceae.

quinine   quinidine

They include **quinine**, **quinidine** and **cinchonidine** which have been used as antimalarial drugs. Quinidine inhibits Na^+-channels and has antiarrhythmic properties. Quinine is very bitter and is employed as a bittering agent in the food industry. It also gives the bitter taste to tonic water, in concentrations of ca. 70 mg per litre.

Peganine (= vasicine) and **vasicinone** (and related compounds) show cholinergic activity. They occur as major alkaloids in *Justicia adhatoda* (an important traditional Ayurvedic medicine) and other members of the Acanthaceae.

peganine (=vasicine)

vasicinone

Most quinoline alkaloids intercalate DNA and thus cause frame shift mutations. Furanoquinolines can be activated by light and can form covalent bonds with DNA bases. This explains their cytotoxicity, antibacterial and antifungal properties.

γ-fagarine

dictamnine

skimmianine

When human skin that has been in contact with furanoquinolines, such as **fagarine**, **dictamnine** or **skimmianine**, is exposed to sunlight, severe burns can occur with blister formation, inflammation and necrosis.

acronine (=acronycine)

rutacridone

Acridone alkaloids have been found in many genera of the Rutaceae. Some of them are potent antineoplastic agents. Examples include **acronine** (= acronycine), extracted from *Acronychia baueri* and **rutacridone**, found in *Ruta graveolens*.

Steroidal alkaloids

Steroidal alkaloids, which often consist of a lipophilic steroid moiety and a hydrophilic oligosaccharide chain, are produced by four unrelated plant families: Apocynaceae, Buxaceae, Liliaceae and Solanaceae. They are especially widely distributed within the very large genus *Solanum* which includes potato, tomato and other food plants. These alkaloids are of the spirosolane type, with glycosides of **soladulcidine** and tomatidine as examples, or of the solanidane type, with glycosides of **solanidine** (e.g. solanine and chaconine) as examples.

soladulcidine

solanidine

Solanum alkaloids behave like saponins (see under saponins). This property also explains the strong skin irritation seen on mucosa and the antibacterial and antifungal properties known from saponins. In addition, the alkaloids inhibit acetylcholine esterase that breaks down acetylcholine in the synapse. Therefore, the *Solanum* alkaloids cause some neuronal effects. Several *Solanum* species, such as *Solanum dulcamara*, are part of traditional medicine used as anti-inflammatory drugs. *Solanum* alkaloids have been used in agriculture as an insecticide.

cyclobuxine D

buxamine E

Plants of the genus *Buxus* (European box and other species) contain a series of free steroidal alkaloids, such as **cyclobuxine D** and **buxamine E**, which are quite toxic and strongly purgative.

Tropane alkaloids

A number of plants, extracts and pure tropane alkaloids have a long history of magic and murder. They have been taken since antiquity to generate hallucinations and intoxication.

Cassava (*Manihot esculenta*): cyanogenic glucosides (linamarin)

Black mustard (*Brassica nigra*): glucosinolates (p. 90)

Coral pea (*Abrus precatorius*): seeds – abrin (toxic lectins)

Castor oil plant (*Ricinus communis*): seeds – ricin (toxic lectins)

Plants containing glucosinolates are often used as spices or vegetables. Mustard oils have been employed in traditional medicine to treat rheumatism (topical application, as counter-irritants) and bacterial infections.

Lectins and peptides

Lectins are small glycosylated and protease-resistant proteins, which are common in seeds of several plants, such as abrin in *Abrus precatorius*, phasin in *Phaseolus vulgaris*, robin in *Robinia pseudoacacia*, and ricin in *Ricinus communis*. Many lectins are highly toxic; ricin is one of the deadliest of all plant poisons. Less toxic lectins occur in seeds of several plants, especially of legumes (Fabaceae) and mistletoe (*Viscum album*), which has been used in phytomedicine. Some of them contribute to allergic properties of a plant, such as peanut lectin (PNA) in peanut seeds (*Arachis hypogaea*) and ragweed pollen allergen (Ra5) from *Ambrosia elatior*. In plants, seed lectins serve as defence compounds against herbivores and nitrogen storage compounds that are remobilised during germination.

Lectins bind to cells via the haptomer (haemagglutinating activity) and become internalised by endocytosis. Once in the cell, they have an affinity for ribosomes and the A-chain (which has N-glycosidase activity) blocks ribosomal protein translation by inactivating elongation factors EF1 and EF2. A cell that is no longer able to make proteins will die.

Lectins are toxic when taken orally, but much more toxic when injected intramuscularly or intravenously. They are among the most toxic peptides produced in nature. Other toxic peptides are found in the venom of snakes, spiders, other animals and in some bacteria (causing whooping cough, cholera or botulism).

Lectins and small peptides can be inactivated by heat; therefore, extensive cooking in water at more than 65 °C usually destroys these toxins. This is why it is important to cook beans and other pulses before they are eaten.

Seeds of several plants accumulate other small peptides such as protease inhibitors. They inhibit the activity of intestinal proteases, such as trypsin and chymotrypsin.

Some plants are rich in hydrolytic proteases, such as bromelain in *Ananas comosus*, ficin in *Ficus glabrata* and papain in *Carica papaya*. They are used medicinally to treat inflammation and digestive problems.

Several small antimicrobial peptides (AMPs) are present in many plants but often overlooked in phytochemical analyses. AMPs exhibit powerful antimicrobial activities.

Format of species monographs, with abbreviations and conventions used

1. **Scientific (botanical) name**
2. **Vernacular name(s) in English**
3. **Top photograph:** main species treated (an alternative source of the drug may be shown as a second photo – then specified in the text)
4. **Classification**
   Type of toxin is indicated, with the World Health Organisation's classification of toxins (see p. 24)
   AHP = included in the African Herbal Pharmacopoeia (2010)
   Clinical studies– = no supportive evidence of efficacy reported after at least one clinical study
   Clinical studies+ = efficacy supported by at least one clinical trial
   Comm. E– = negative monograph by the German Commission E
   Comm. E+ = positive monograph by the German Commission E
   DS = Dietary Supplement
   ESCOP = included in ESCOP monographs (European Scientific Cooperative on Phytotherapy)
   ESCOP Suppl. = included in later supplements of the ESCOP monographs
   HMPC = included in the monographs of the European Medicines Agency (Herbal Medicinal Products)
   MM = Modern Medicine
   Pharm. = included in one or more pharmacopoeia
   PhEu8 = included in the 8th edition of the European Pharmacopoeia (2014)
   TCM = Traditional Chinese Medicine
   TM = Traditional Medicine
   WHO 1 = Monographs of the World Health Organisation (plus volume number)
5. **Uses & properties**
   Pharmaceutical name is provided (in English and Latin)
   The plant part used is specified (see p. 28)
   Dosage forms (see p. 30) and methods of administration (see p. 34)
   Main uses are indicated, and the dose where available (1 g = 0.035274 oz)
6. **Origin**
   Geographical region (usually continent) where the plant is indigenous
   Region(s) where it is naturalised and/or cultivated is sometimes given
7. **Botany**
   A very brief description, with emphasis on growth form (habit)
   Botanical terms are defined in the glossary (p. 274)
8. **Chemistry**
   A brief summary of the main chemical compounds of pharmacological and/or toxicological interest
   The name printed in bold refers to the given chemical structure (often the main active ingredient)
   Yields of compounds are given as % of dry weight (unless specified otherwise)
   A brief introduction to the various classes of compounds is presented elsewhere (see p. 49)
9. **Pharmacology**
   A brief summary of preclinical and clinical data, with biological activities of main compound(s)
   For definitions of pharmacological terms see glossary (p. 274)
10. **Toxicology**
    An indication of toxicity, safety, edibility and possible side effects
    LD_{50} = the dose of substance (in mg per kg body weight) that kills 50% of the specified test animals
    Methods of administration: p.o. = per os (ingested by mouth); i.p. = intraperitoneal injection; i.v. = intravenous injection; s.c. = subcutaneous injection
11. **Bottom photograph:** usually the fresh or mostly the dried raw material, as it is traded
    In the case of poisonous plants, other details (or related species) are shown
12. **Last paragraph (below photo):** scientific name of the species, with author citation, main synonym and family name; vernacular name(s) in several languages (space permitting)

Abrus precatorius
crab's eye vine • coral pea

CLASSIFICATION Cell toxin, extremely hazardous (Ia); TM: Africa, Asia.
USES & PROPERTIES The attractive seeds are used to make necklaces, rosaries, bracelets and other decorative objects. A highly resistant seed coat ensures that the intact seeds pass harmlessly through the digestive tract. However, when seeds are pierced (to make beads) or damaged, the poison is released, causing dermatitis, intoxication and even death.
ORIGIN Africa, Asia.
BOTANY Woody climber; leaves pinnate; flowers pale purple; pods 4–5-seeded; seeds 5 mm in diam.
CHEMISTRY **Abrin** (a mixture of four lectins, called abrin A–D, are present in seeds); abrusosides (sweet-tasting triterpene saponins, occur in the leaves and roots).
PHARMACOLOGY Abrin: haemagglutinating, inhibitor of ribosomal protein synthesis.
TOXICOLOGY Abrin: LD_{50} = 0.02 mg/kg (mouse, i.p.); seeds: lethal dose = 0.5 g (humans, p.o.).
NOTES Fatal cases of poisoning are rare.

abrin (lectin)

Abrus precatorius L. (Fabaceae); *pois rouge* (French); *Paternostererbse* (German)

Acacia senegal
gum acacia • gum arabic tree

CLASSIFICATION TM: Africa, Europe, Asia. Pharm., PhEur8, AHP.
USES & PROPERTIES Gum arabic is the tasteless and odourless dried exudate collected from the bark. It is used topically as emollient to promote healing and to protect the skin and mucosa from bacterial and fungal infections. The main use in pharmacy is as emulsifier, stabiliser of suspensions and additive for solid formulations and tablets.
ORIGIN Africa. Gum is produced in North Africa and especially in Sudan and Ethiopia.
BOTANY Deciduous tree (to 6 m); thorns typically in groups of three; leaves pinnately compound; flowers minute, cream-coloured, in elongated spikes; pods flat, oblong.
CHEMISTRY Gum arabic is a polysaccharide (MW 270 000) with **arabinose**, galactose, D-glucuronic acid and L-rhamnose subunits.
PHARMACOLOGY Moisturising, antibiotic and protective effects on skin and mucosa.
TOXICOLOGY Non-toxic (edible).

Acacia senegal (L.) Willd. (Fabaceae); *acacie gomme arabique* (French); *Verek-Akazie* (German); *acacia del Senegal* (Italian)

Achillea millefolium
yarrow • milfoil • woundwort

CLASSIFICATION TM: Asia, Europe. Pharm., Comm. E+, ESCOP Suppl., PhEur8, HMPC.
USES & PROPERTIES The whole plant (*Millefolii herba*), flowers (*Millefolii flos*) or sometimes the essential oil are used for lack of appetite and minor dyspeptic complaints. Traditional uses include the treatment of arthritis, the common cold, fever and hypertension. Internal use: 4.5 g of the herb per day, as infusion or tincture (or 3 g flowers). External use: 100 g herb in 20 litres of bath water.
ORIGIN Europe and W Asia (widely cultivated).
BOTANY Perennial herb; leaves compound, feathery; flowers white to pink.
CHEMISTRY Pyrrolidine alkaloids (betonicine, stachydrine) flavonoids and essential oil (α-pinene, camphor, 1,8-cineole, caryophyllene and blue azulenic compounds released from lactones (e.g. **achillicin**) during steam distillation).
PHARMACOLOGY Antibacterial, anti-inflammatory, antispasmodic; antipyretic, hypotensive.
TOXICOLOGY Low toxicity; may cause dermatitis.

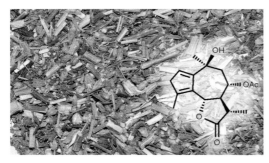

Achillea millefolium L. (Asteraceae); *millefeuille* (French); *Schafgarbe* (German); *achillea millefoglio* (Italian); *milenrama* (Spanish)

Aconitum napellus
aconite • monkshood • wolfsbane

CLASSIFICATION Neurotoxin, mind-altering (Ia). TM: Europe, Asia. MM and homoeopathy.
USES & PROPERTIES Dilute root tinctures are used in cough syrups and in homoeopathy. Higher concentrations (or pure alkaloid) are applied topically to treat rheumatism and neuralgia. Aconite is a psychoactive drug. In India and China, some species are used topically for analgesic, antineuralgic, anti-inflammatory and antipyretic effects. Formerly used for executions, murder, suicide and to control vermin (hence "wolfsbane").
ORIGIN Europe (widely cultivated).
BOTANY Perennial herb; root tuberous; leaves dissected; flowers hood-shaped, sepals colourful.
CHEMISTRY Diterpene alkaloids (**aconitine**).
PHARMACOLOGY Aconitine stimulates Na^+-channels; peripheral nerve endings are first activated and then paralysed. It is strongly psychedelic when smoked or absorbed through the skin.
TOXICOLOGY Aconitine: lethal dose 3–6 mg (humans). All *Aconitum* species are very toxic.

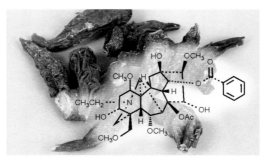

Aconitum napellus L. (Ranunculaceae); *aconit napel* (French); *Blauer Eisenhut* (German); *aconito* (Italian); *acónito* (Spanish)

Acorus calamus
calamus • sweet flag

CLASSIFICATION Cell toxin, mutagenic. TM: Asia (Ayurvedic and Chinese medicine). Pharm.

USES & PROPERTIES Fresh or dried rhizomes are traditionally used as aromatic bitter tonics and appetite stimulants to treat indigestion, flatulence, stomach cramps, chronic dysentery and asthma.

ORIGIN North temperate zone (Asia, Europe and North America). It grows in water or wet places.

BOTANY Perennial herb (0.8 m); leaves linear, midribs distinct; flowers minute, in oblong spikes.

CHEMISTRY Essential oil: sesquiterpenoids (**acorenone**) and several monoterpenoids (e.g. camphene, *p*-cymene, linalool). Phenylpropanoids (β-asarone) in the Indian variety.

PHARMACOLOGY Spasmolytic, CNS sedative, toxic and mutagenic properties are ascribed to β-asarone. It is a potent carcinogen in rodents that may induce duodenal and liver cancer.

TOXICOLOGY β-asarone: LD_{50} = 184 mg/kg (mouse, i.p.). In Europe, a maximum of 0.1 mg/kg is allowed in foodstuffs. *Acorus* and its oil is prohibited for food use in the USA.

Acorus calamus L. (Acoraceae); *acore vrai* (French); *bacc* (Hindi); *Kalmus* (German); *calamo aromatico* (Italian); *cálamo aromático* (Spanish)

Adansonia digitata
baobab

CLASSIFICATION TM: Africa. AHP. DS: fruit.

USES & PROPERTIES The fruits and seeds are traditionally used in African traditional medicine to treat dysentery and fever. Numerous medicinal uses have also been reported for the leaves, bark and roots. The dry and powdered fruit pulp has become popular as a dietary supplement and health food item: in health drinks such as smoothies, fruit juices (6–8%) and cereal health bars (5–10%).

ORIGIN Africa. Trees are cultivated on a small scale in warm, tropical regions.

BOTANY Massive deciduous tree (to 15 m); trunk 20 m or more in circumference; leaves digitate; flowers large, white; fruit ovoid, velvety.

CHEMISTRY Fruit pulp contains ascorbic acid, **tartaric acid** and water-soluble pectins (to 56%).

PHARMACOLOGY Antioxidant activity (comparable to that of grape seed extract). Demonstrated activities (at high doses): analgesic, anti-inflammatory and antipyretic.

TOXICOLOGY Fruit pulp is safe to consume. It has Novel Food (EU) and GRAS (USA) status.

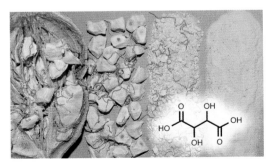

Adansonia digitata L. [Malvaceae (formerly Bombacaceae)]; *adansonie d'Afrique* (French); *Affenbrotbaum* (German); *baobab africano* (Italian); *baobab del África* (Spanish)

Adenium obesum
desert rose

CLASSIFICATION Heart poison (Ia).
USES & PROPERTIES The stems and roots contain watery latex that has been used as fish poison and arrow poison, often in combination with other toxic plants and plant parts.
ORIGIN Africa (southern and eastern parts).
BOTANY Thick-stemmed woody shrub (to 3 m); leaves glossy green, deciduous; flowers showy, shades of red, pink and white; fruits cylindrical, dehiscent capsules; seeds numerous, hairy.
CHEMISTRY More than 30 cardiac glycosides (**obebioside B** is the main compound).
PHARMACOLOGY Cardiac glycosides: inhibition of the crucial Na^+, K^+-ATPase ion pump, which results in slow heartbeat, arrhythmia, ventricular fibrillation, cardiac arrest and death.
TOXICOLOGY Poisoned arrows may cause death within a few minutes. Lethal dose unknown (perhaps 0.1–0.6 mg/kg body weight (humans, i.v.); more for oral toxicity.
NOTES *Adenium multiflorum* (impala lily), also known as *A. obesum* var. *multiflorum*, is shown below.

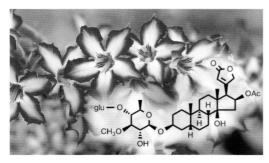

Adenium obesum Forskal [= *A. obesum* var. *obesum*] (Apocynaceae); *rose du désert* (French); *Wüstenrose* (German)

Adonis vernalis
yellow pheasant's eye • spring adonis

CLASSIFICATION Heart poison (Ib). TM: Europe. Pharm., Comm. E+, homoeopathy.
USES & PROPERTIES Aerial flowering parts (*Adonidis herba*), in carefully controlled doses of a standardised mixture (see below), to treat the symptoms of heart insufficiency. Traditional use: bladder and kidney stones.
ORIGIN Europe (excluding Britain), Siberia.
BOTANY Perennial herb; leaves compound, feathery; flowers large, bright yellow.
CHEMISTRY More than 20 cardenolides (0.2 to 0.5%); **cymarin** as main compound (hydrolysis gives kappa-strophanthidin and *D*-cymarose).
PHARMACOLOGY Cymarin: positive inotropic and venotonic activity; strengthen the contraction of the heart muscle without increasing the pulse.
TOXICOLOGY Cymarin: LD_{50} = 0.11–0.13 mg/kg (cat, i.v.); average daily dose of standardised powder (with activity equivalent of 0.2% cymarin: 0.6 g; max. single dose: 1 g, max. daily dose: 3 g.
NOTES Warning: use only under the supervision of a qualified health care professional.

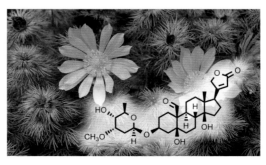

Adonis vernalis L. (Ranunculaceae); *adonide du printemps* (French); *Frühlings-Adonisröschen* (German); *adonide* (Italian); *botón de oro* (Spanish)

Aesculus hippocastanum
horse chestnut

CLASSIFICATION Cell toxin (III). TM: Europe. Comm. E+, ESCOP 6, WHO 2, HMPC. MM (aescin). Clinical studies+.

USES & PROPERTIES Standardised extracts of dried seeds (*Hippocastani semen*) are used in modern phytotherapy to treat the symptoms of venous and lymphatic insufficiency. Leaves (*Hippocastani folium*): in TM (coughs, arthritis and rheumatism; in galenical preparations to treat venous conditions). Bark (*Hippocastani cortex*): rarely in TM (diarrhoea and haemorrhoids).

ORIGIN Eastern Europe to Central Asia.

BOTANY Large tree (to 30 m); leaves digitate; flowers showy, white with pink spots; fruit a spiny capsule; seeds large, smooth, brown.

CHEMISTRY Aescin (a mixture of triterpene saponins, to 5%; 16–21% in standardised extracts); main compound: a glycoside of **protoaescigenin**.

PHARMACOLOGY Anti-inflammatory, venotonic and anti-oedema (increased vascular tone and stability of capillary veins).

TOXICOLOGY Moderately toxic.

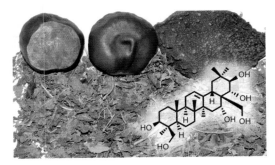

Aesculus hippocastanum L. [Sapindaceae (formerly Hippocastanaceae)]; *marronnier d'Inde* (French); *Rosskastanie* (German); *castagna amare* (Italian); *castaño de Indias* (Spanish)

Aframomum melegueta
Melegueta pepper • grains of paradise

CLASSIFICATION TM: Africa. AHP.

USES & PROPERTIES The seeds are used in traditional medicine to treat a wide range of ailments including infertility, cough, measles, leprosy, dysentery, stomach ailments and indigestion.

ORIGIN Africa (tropical western parts).

BOTANY Perennial herb (to 1 m) with creeping rhizomes; leaves large, glabrous; flowers white or pink, showy; fruit ovoid, fleshy, bright red; seeds many, angular, reddish brown, tuberculate/warty.

CHEMISTRY The seeds contain **[6]-paradol**, [6]-shogaol and other hydroxyalkylalkanones. The essential oil has α-humulene and β-caryophyllene as main compounds.

PHARMACOLOGY Both [6]-paradol and the essential oil have demonstrated antimicrobial activity. Extracts of the seeds have aphrodisiac, anti-inflammatory and analgesic properties.

TOXICOLOGY The seeds are edible and have no toxic effects even at doses of 4 g/kg (rat, p.o.).

NOTES Once an important substitute for black pepper and still a valuable spice.

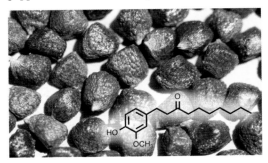

Aframomum melegueta (Roscoe) K. Schum. (Zingiberaceae); *poivre de Guinée* (French); *Malagettapfeffer, Paradieskörner,* (German); *grani de melegueta* (Italian); *malagueta* (Spanish)

Agathosma betulina
buchu • round leaf buchu

CLASSIFICATION TM: Africa (Cape). Pharm., Comm.E–, AHP.
USES & PROPERTIES Fresh or dried leaves (*Barosmae folium; Folia Bucco*) are traditionally used as diuretic, diaphoretic and stimulant tonic (kidney and bladder ailments, rheumatism, and minor digestive disturbances). In the form of "buchu vinegar", it has been applied to wounds and bruises.
ORIGIN Africa (Western Cape, South Africa).
BOTANY Woody shrublet; leaves gland-dotted, rounded (oval – more than twice as broad) in the related oval leaf buchu (*A. crenulata*); flowers white to pale purplish; fruit a 5-seeded capsule.
CHEMISTRY Essential oil: rich in **diosphenol** (buchu camphor) but also limonene, isomenthone and terpinen-4-ol as main compounds. Leaves contain flavonoids (mainly diosmin).
PHARMACOLOGY Buchu leaf and buchu oil are believed to have urinary antiseptic, diuretic and anti-inflammatory activity.
TOXICOLOGY Round leaf buchu is non-toxic but oval leaf buchu oil has pulegone (potentially toxic).

Agathosma betulina (Berg.) Pillans [= *Barosma betulina* (Berg.) Bartl. & H.L. Wendl.] (Rutaceae); *buchu* (French); *Bucco* (German); *buchu* (Italian)

Agrimonia eupatoria
common agrimony

CLASSIFICATION TM: Europe, Asia. Pharm., Comm.E+, ESCOP, PhEur8.
USES & PROPERTIES Dried aboveground parts (*Agrimoniae herba*) are used as antidiarrhoeal, astringent and mild diuretic for treating throat infections, acute diarrhoea, cystitis, piles and ailments of the urinary tract. It is a traditional styptic to treat bleeding wounds and is claimed to have value in alleviating the symptoms of arthritis and rheumatism. An ingredient of mixtures to treat disorders of the stomach, liver and gall bladder.
ORIGIN Europe and the Near East.
BOTANY Erect perennial herb; leaves compound, toothed, hairy; flowers small, yellow, in slender spikes. Fragrant agrimony (*A. procera*) and Chinese agrimony (*A. pilosa*) are also used.
CHEMISTRY **Catechin** tannins and gallotannins (agrimoniin is one of the main compounds).
PHARMACOLOGY Tannins are astringent and bind to proteins (hence the antiviral and antibacterial activity). Diuretic activity (not proven).
TOXICOLOGY Very low toxicity.

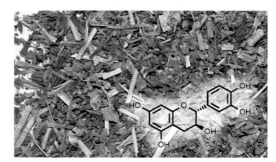

Agrimonia eupatoria L. (Rosaceae); *aigremoine gariot* (French); *Kleiner Odermennig* (German); *agrimonia* (Italian); *agrimonia* (Spanish)

Agrostemma githago
corn cockle

CLASSIFICATION Cell toxin (Ib). TM: Europe.
USES & PROPERTIES The plant was once an abundant weed in cereal fields and the seeds a common contaminant of grain, resulting in human and animal poisoning (sometimes with fatal results). Used to a limited extent in traditional medicine: cough, gastritis and skin ailments.
ORIGIN Europe (Mediterranean region; an introduced weed elsewhere).
BOTANY Erect annual; leaves linear-lanceolate; flowers purple (sepals longer than the petals); capsules many-seeded; seeds orbicular, black.
CHEMISTRY Toxic triterpene saponins, especially githagin; the aglycone is **githagenin** (=gypsogenin); a toxic lectin: agrostin.
PHARMACOLOGY Saponins increase the permeability of biomembranes and thus the uptake of the lectin (the latter inhibits protein synthesis).
TOXICOLOGY Saponins: LD_{50} = 2.3 mg/kg (rats, i.p.), 50 mg/kg (rats, p.o.); seeds: lethal dose = 5 g or more (humans, p.o.).
NOTES Nowadays, cases of poisoning are rare.

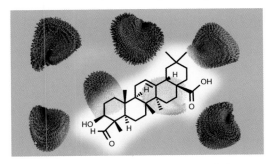

Agrostemma githago L. (Caryophyllaceae); *nielle des champs* (French); *Kornrade* (German); *gittaione* (Italian)

Alchemilla xanthochlora
lady's mantle

CLASSIFICATION TM: Europe. Pharm., Comm. E+, PhEur8.
USES & PROPERTIES The dried aerial parts of flowering plants (*Alchemillae herba*) are mainly used to treat mild diarrhoea (daily dose of 5–10 g) and throat infections. Externally it is applied to wounds and sores. Traditional uses include the treatment of dysmenorrhoea. The dry product (or extracts) is included in herbal mixtures (for diarrhoea) and in mouthwashes and lozenges (for sore throat).
ORIGIN Europe, North America, Asia.
BOTANY Low perennial herb; leaves lobed, toothed; flowers small, yellow, in sparse clusters.
CHEMISTRY **Ellagitannins** (up to 8%): agrimoniin and others.
PHARMACOLOGY Tannins are highly astringent, non-specific protein poisons. They precipitate the proteins of the bacteria that cause diarrhoea and sore throat.
TOXICOLOGY Lady's mantle is non-toxic at the recommended daily dose.

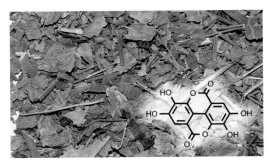

Alchemilla xanthochlora Rothm. [= *Alchemilla vulgaris* auct. non L.] (Rosaceae); *alchimille* (French); *Gewöhnlicher Frauenmantel* (German); *alchemilla* (Italian); *pie de leon* (Spanish)

Aleurites fordii
tung oil tree

CLASSIFICATION   Cell toxin, skin irritant (II).
USES & PROPERTIES   Ripe seeds are the source of commercial tung oil, an ingredient of paints and quick-drying varnishes. Poisoning may occur if the fruits or seeds are eaten or if the oil is ingested.
ORIGIN   Asia (China).
BOTANY   Tall tree (12 m); branches thick; leaves heart-shaped (two small glands are present at the insertion of the long petiole); flowers white, marked with pink or red; Capsules large, pendulous, few-seeded. The species is sometimes included in the genus *Vernicia* (as *V. fordii*).
CHEMISTRY   Diterpenoids: **phorbol** esters of the tigliane type (the main compound is 12-*O*-palmitoyl-13-*O*-acetyl-16-hydroxyphorbol).
PHARMACOLOGY   Phorbol esters are not only strongly purgative but potent tumour promoting agents (co-carcinogens: activation of cell division). They are skin irritants and cause painful inflammation of mucous membranes (including the eye).
TOXICOLOGY   Seeds cause vomiting, stomach pain and diarrhoea; recovery within 1–2 days.

Aleurites fordii Hemsl. [= *Vernicia fordii* (Hemsl.) Airy Shaw] (Euphorbiaceae); *alévrite*, *bois de Chine* (French); *Tungölbaum* (German)

Allium cepa
onion

CLASSIFICATION   TM (DS): Europe, Asia. Comm.E+, WHO 1, HMPC, clinical studies+.
USES & PROPERTIES   The fresh or dried bulb (*Allii cepae bulbus*) is used as antibiotic, for cholesterol-lowering activity and to treat appetite loss. Doses of 50 g fresh onion (20 g dry) are considered effective in treating or preventing arteriosclerosis (age-related changes in blood vessels) and blood-clotting. Many traditional uses: minor digestive disturbances, insect stings, wounds, colds, cough, asthma, dysentery and diabetes.
ORIGIN   Uncertain (cultigen, possibly from the eastern Mediterranean region).
BOTANY   Perennial herb; leaf bases forming a bulb; flowers small, white, in a rounded head.
CHEMISTRY   Sulfur-containing compounds: sulfoxides (especially isoalliin) in the intact bulb (converted to **isoallicin** when cells are damaged).
PHARMACOLOGY   Antimicrobial, hypoglycaemic, anti-platelet aggregation, anti-asthmatic, anti-allergic, lipid-lowering, hypotensive.
TOXICOLOGY   Onions are non-toxic (edible).

Allium cepa L. [Amaryllidaceae (formerly Alliaceae)]; *oignon* (French); *Küchenzwiebel* (German); *cipolla* (Italian); *cebolla* (Spanish)

Allium sativum
garlic

CLASSIFICATION TM (DS): Europe, Asia. Comm.E+, ESCOP 3, WHO 1, PhEur8, HMPC. Clinical studies+.

USES & PROPERTIES Fresh bulb segments (*Allii sativi bulbus*), garlic powder (*Allii sativi pulvus*) or garlic oil are used for their lipid-lowering effects and also for antibacterial and antiviral activity. A daily dose of 4 g in fresh garlic (or equivalent in powder or oil) is used as supportive treatment for high blood cholesterol and the common cold.

ORIGIN Uncertain (cultigen, from Central Asia?).

BOTANY Perennial herb arising from a subterranean bulb comprising segments or cloves (axillary buds); leaves flat; flowers small, white, in a small cluster enclosed in a sheath-like bract.

CHEMISTRY Sulfur-containing compounds: sulfoxides (mainly alliin) in intact bulbs; converted to **allicin** (major compound in garlic products).

PHARMACOLOGY Demonstrated antimicrobial, antiviral, lipid-lowering and platelet aggregation inhibiting activities.

TOXICOLOGY Edible (interaction with warfarin).

Allium sativum L. [Amaryllidaceae formerly Alliaceae)]; *ail blanc* (French); *Knoblauch* (German); *aglio* (Italian); *ajo* (Spanish)

Allium ursinum
wild garlic • bear's garlic • ramsons

CLASSIFICATION TM (DS): Europe, Asia.

USES & PROPERTIES Fresh leaves (*Allii ursini herba*) or fresh bulbs (*Allii ursini bulbus*) are used in the same way as garlic (*A. sativum*). In traditional medicine: to treat digestive ailments and arteriosclerosis; externally to alleviate skin allergies. The leaves (often used as spring salad) have become popular dietary supplement or functional food and are an ingredient of commercial mixtures.

ORIGIN Europe.

BOTANY Bulbous perennial herb; leaves broad, short, flat; flowers small, white, in sparse clusters.

CHEMISTRY Sulfur-containing compounds (up to 12% in fresh bulbs): sulfoxides (mainly alliin and **methiin**) are enzymatically converted upon drying or processing.

PHARMACOLOGY As with garlic, antimicrobial, antiviral and lipid-lowering effects can be expected.

TOXICOLOGY Bear's garlic is edible but care should be taken with wild-harvested material: the leaves are easily confused with those of the poisonous lily-of-the-valley (*Convallaria majalis*).

Allium ursinum L. [Amaryllidaceae (formerly Alliaceae)]; *ail des ours* (French); *Bärlauch* (German); *aglio orsino* (Italian); *ajo de oso* (Spanish)

Aloe ferox
bitter aloe • Cape aloe

CLASSIFICATION TM (DS): Africa. Comm.E+, ESCOP 5, WHO 2, PhEur8, HMPC, AHP.
USES & PROPERTIES The bitter yellow leaf exudate is dried to a dark brown solid (commercial aloe lump or Cape aloes; *Aloe capensis*). It is a laxative medicine but also an ingredient of bitter tonic drinks. The non-bitter inner gel is used in tonic drinks and cosmetic preparations.
ORIGIN Africa (SE parts of South Africa).
BOTANY Single-stemmed succulent (2 m); leaves with sharp teeth; flowers in erect racemes.
CHEMISTRY Leaf exudate: anthraquinones, mainly **aloin** (= barbaloin), an anthrone-*C*-glycoside present as isomers (aloin A and aloin B). Leaf gel: polysaccharides and glycoproteins.
PHARMACOLOGY Aloin: a prodrug that is converted in the colon (by bacterial action) to aloe-emodin anthrone. The latter has stimulant laxative activity. Gel: hydrating and insulating effects (wound-healing and soothing properties).
TOXICOLOGY Anthraquinones may be carcinogenic (avoid chronic use or use during pregnancy).

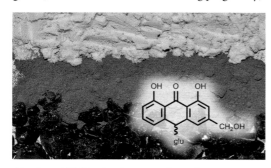

Aloe ferox Mill. [Xanthorrhoeaceae (formerly Asphodelaceae)]; *aloès féroce* (French); *Kap-Aloe, Gefährliche Aloe* (German); *aloe del Capo* (Italian)

Aloe vera
aloe vera • Curaçao aloe

CLASSIFICATION TM (DS): Africa, Asia. Comm.E+, WHO 1 (gel), PhEur8.
USES & PROPERTIES Non-bitter leaf parenchyma (obtained by manual or mechanical "filleting") is the main product, widely used as tonic drinks but also in skincare and cosmetic products. The bitter leaf exudate (Barbados aloes, Curaçao aloes) is less often used: as stimulant laxative and bitter tonic.
ORIGIN North Africa or Arabia (ancient cultigen).
BOTANY Stemless succulent; leaves fleshy, with harmless marginal teeth; flowers usually yellow.
CHEMISTRY Gel: 0.5–2% solids (polysaccharides – **acemannan**, glycoproteins). Leaf exudate: aloin (= barbaloin), an anthrone-*C*-glycoside (to 38%). Gel drinks usually have < 10 ppm aloin.
PHARMACOLOGY The gel has anti-inflammatory, wound-healing and immune-stimulatory effects (not scientifically fully proven). Aloin is a stimulant laxative with bitter tonic (*amarum*) effects.
TOXICOLOGY Anthraquinones are potentially carcinogenic and long-term use (and use during pregnancy) should be avoided.

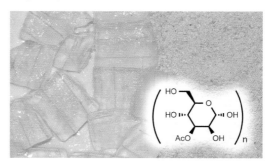

Aloe vera (L.) Burm.f. [= *Aloe barbadensis* Mill.] [Xanthorrhoeaceae (formerly Asphodelaceae)]; *aloès vrai, laloi* (French); *Echte Aloe* (German); *aloe vera* (Italian); *sábila, zábila* (Spanish)

Aloysia citrodora
lemon verbena • vervain

CLASSIFICATION TM (DS): South America, Europe. PhEur8.

USES & PROPERTIES The fresh or dried leaves (*Lippiae triphyllae folium*) are used as a calming health tea (especially in France) to treat digestive ailments and minor sleeplessness. Included in mixtures to treat nervous conditions, colds, fevers, indigestion, flatulence and mild diarrhoea. Essential oil (*Lippiae triphyllae aetheroleum*) is popular in aromatherapy for its pleasant lemon smell.

ORIGIN South America (Argentina and Chile), now often cultivated (especially France and Spain).

BOTANY Woody shrub (to 3 m); leaves in groups of three or four, rough-textured, aromatic (lemon smell); flowers pale mauve, in loose panicles.

CHEMISTRY Essential oil with citral (i.e., a mixture of neral and geranial), **limonene** and several other compounds. Numerous flavonoids are present.

PHARMACOLOGY Sedative and anxiolytic properties are not yet substantiated by clinical studies.

TOXICOLOGY The leaves are often used as herb (flavour ingredient in food and beverages).

Aloysia citrodora Palau [= *A. triphylla* (L'Herit.) Britton; *Lippia citriodora* H.B.K.] (Verbenaceae); *verveine odorante* (French); *Zitronenstrauch* (German); *limoncina* (Italian)

Alpinia officinarum
galangal • Siamese ginger • lesser galangal

CLASSIFICATION TM: China, India, Europe. Pharm., Comm.E+.

USES & PROPERTIES Fresh or dried rhizomes (*Galangae rhizoma*) are used as infusions, tinctures, decoctions or powders for dyspepsia and appetite loss. In China it is also used against indigestion and stomach pain, hiccups and nausea. Tea is made from 0.5–1 g of dried rhizome, taken half an hour before meals (daily dose 2–4 g dry weight).

ORIGIN Eastern and Southeastern Asia; cultivated in China, India, Malaysia and Thailand.

BOTANY Leafy perennial herb (ca. 1 m) with fleshy rhizomes; Flowers white and reddish purple.

CHEMISTRY Essential oil (0.5–1%); non-volatile diarylheptanoids (galangols) and phenyl alkyl ketones (gingerols); flavonoids (**galangin**).

PHARMACOLOGY Galangin has antispasmodic, anti-inflammatory and antibiotic activities; diarylheptanoids inhibit prostaglandin biosynthesis and induce apoptosis in cancer cells (*in vitro*).

TOXICOLOGY The rhizomes are not toxic.

NOTES Greater galangal (*Alpinia galanga*) is a spice.

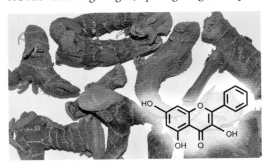

Alpinia officinarum Hance (Zingiberaceae); *gao liang jiang* (Chinese); *galanga* (French); *Echter Galgant* (German); *galanga* (Italian); *galanga* (Spanish)

Althaea officinalis
marshmallow • white mallow

CLASSIFICATION TM: Europe. Comm.E+, ESCOP 1, WHO 2, 5, PhEur8, HMPC.
USES & PROPERTIES The roots (*Althaeae radix*) are used as anti-irritant and expectorant medicine to alleviate the symptoms of cough, peptic ulcers and inflammation of the mouth, throat and stomach. Leaf infusions or marshmallow syrup (*Sirupus Althaeae*) are used to treat dry cough. Preparations are applied to burns, sores and ulcers. Daily dose (sipped or gargled): 6 g root, 5 g leaf (or 10 g marshmallow syrup, as single dose).
ORIGIN Asia; naturalised in America.
BOTANY Erect perennial herb (2 m); leaves hairy; flowers large, pink. Flowers of the related *Alcea rosea* (hollyhock) are similarly used.
CHEMISTRY Mucilages (polysaccharides): composed of **galacturonic acid**, glucuronic acid, galactose, arabinose and rhamnose (up to 15% in roots, less than 10% in leaves and flowers).
PHARMACOLOGY Polysaccharides have a soothing effect on mucosa.
TOXICOLOGY Marshmallow is safe to ingest.

Althaea officinalis L. (Malvaceae); *guimauve* (French); *Eibisch* (German); *bismalva, altea* (Italian); *malvavisco* (Spanish)

Ammi majus
bishop's weed • lace flower

CLASSIFICATION MM (furanocoumarins): photochemotherapy (PUVA).
USES & PROPERTIES The small fruits are a commercial source of phototoxic furanocoumarins that are used to treat skin disorders such as psoriasis and vitiligo.
ORIGIN Europe, Asia and North Africa.
BOTANY Erect annual (0.5 m); leaves compound, with broad, serrate leaflets; flowers 50–100.
CHEMISTRY Linear furanocoumarins: **bergapten** (= 5-methoxypsoralen) and xanthotoxin are major compounds in the fruits ("seeds").
PHARMACOLOGY Furanocoumarins are phototoxic – after ingestion or skin contact, they interact with sunlight or UV light and cause acute dermatitis with blisters and vesicles. This property is used in photochemotherapy or so-called PUVA: oral administration of 20–40 mg furanocoumarins (e.g. psoralen), followed (2 hours later) by exposure to sunlight/UVA radiation (ca. 20 sessions needed).
TOXICOLOGY PUVA increases the risk of skin and lung cancer.

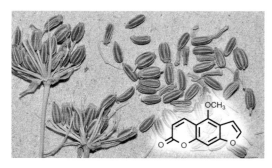

Ammi majus L. (Apiaceae); *grand ammi* (French); *Große Knorpelmöhre* (German); *visnaga maggiore* (Italian); *âmio-maior, âmio-vulgar* (Portuguese); *ameo mayor* (Spanish)

Anadenanthera peregrina
cohoba • yopo • niopo

CLASSIFICATION Hallucinogen, mind-altering, toxin (II).
USES & PROPERTIES The bark and seeds are used as psychotropic drug that has powerful hallucinogenic effects. It is a so-called entheogen used for centuries in religious, shamanic and spiritual ceremonies and rituals. Seed powder, in combination with alkaline ash, is most commonly used in an elaborate process to make preparations that are taken as snuff or as enema. It can also be taken orally (mixed with maize beer or honey) or smoked (with tobacco). The average dose is 5–10 g.
ORIGIN South America (Brazil, Andes).
BOTANY Tree (to 20 m); leaves pinnate; flowers white; pods 6–15-seeded; seeds thin, black.
CHEMISTRY Tryptamine derivatives: the alkaloid **bufotenin** is mainly responsible for the activity.
PHARMACOLOGY Bufotenin and tryptamine derivatives activate serotonin receptors.
TOXICOLOGY Yopo has a low toxicity. The LD_{50} = 200–300 mg/kg (rat, i.p.) but there are unpleasant side effects.

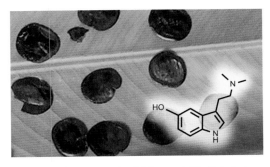

Anadenanthera peregrina (L.) Spegazzini [= *Piptadenia peregrina*] (Fabaceae); *yopo* (French); *Yopo* (German)

Ananas comosus
pineapple

CLASSIFICATION MM (DS): Europe. Comm.E+.
USES & PROPERTIES Pineapple juice is traditionally taken for its digestive tonic and diuretic effects. A mixture of proteolytic enzymes, extracted from fruits and stems, is known as bromelain (*Bromelainum crudum*). It is used to treat post-traumatic and post-operative swellings (and to alleviate digestive disorders). Daily dose: 80–240 mg (as tablets, 2–3 times per day, up to 10 days).
ORIGIN Central America; widely cultivated as a fruit crop in tropical parts of Africa and Asia.
BOTANY Perennial herb; leaves tough, fibrous, in rosettes, margins spiny; flower cluster purple, all parts becoming fleshy as the fruit develops.
CHEMISTRY Bromelain: a mixture of five or more proteolytic enzymes (mainly bromelain A and B). Rich in vitamin C (20 mg per 100 g of ripe fruit).
PHARMACOLOGY Bromelain: demonstrated anti-inflammatory, anti-oedemic, anti-platelet and fibrinolytic activities.
TOXICOLOGY Side effects of bromelain: allergic reactions, diarrhoea and indigestion.

bromelain (mixture of proteolytic enzymes)

Ananas comosus (L.) Merr. [= *A. sativa*] (Bromeliaceae); *ananas* (French); *Ananas* (German); *ananasso* (Italian); *piña* (Spanish)

Andrographis paniculata
king of bitters • kalmegh

CLASSIFICATION TM: Ayurveda, Siddha, TCM. WHO 2, HMPC.

USES & PROPERTIES Aboveground parts (*Andrographidis paniculatae herb*a) are used, in doses of 3–9 g per day, for an exceptionally wide range of ailments. It is a traditional bitter tonic and adaptogen, often used to treat fever, colds, flu, digestive complaints, sinusitis, bronchitis and tonsillitis.

ORIGIN India and Sri Lanka (widely cultivated).

BOTANY Annual herb (1 m); leaves lanceolate; flowers two-lipped, white with purple marks.

CHEMISTRY Diterpene lactones (of the labdane type, up to 6%): **andrographolide** (the main bioactive compound) and several other diterpenes.

PHARMACOLOGY Anti-inflammatory: clinical evidence of efficacy in the treatment and prevention of upper respiratory tract infections. Reported analgesic and antimicrobial activity. Andrographolide is intensely bitter and is probably responsible for most of the claimed health benefits.

TOXICOLOGY Very low toxicity. Whole herb: LD_{50} = 40 g/kg (mouse, p.o).

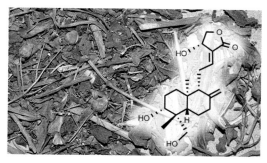

Andrographis paniculata (Burm. f.) Nees. [= *Justicia paniculata* Burm. f.] (Acanthaceae); *chuan xin lian* (Chinese); *Kalmegh* (German); *kirayat* (Hindi); *bhunimba* (Sanskrit)

Anethum graveolens
dill

CLASSIFICATION TM: Europe; Pharm., Comm. E+ (fruits only).

USES & PROPERTIES The fruits (*Anethi fructus*) are traditionally used as diuretic and to treat digestive disorders (especially dyspepsia and flatulence). The daily dose is ca. 3 g (as infusion or tincture). Essential oil of the fruits (*Anethi aetheroleum*) is similarly used (daily dose: 0.1–0.3 g). The fruits are also an ingredient of gripe water for treating colic and flatulence in infants. Dill herb (*Anethi herba*) is best known for its culinary value.

ORIGIN Probably SW Asia (but cultivated since ancient times in Egypt, Europe and Asia).

BOTANY Slender annual herb; leaves feathery; flowers small, yellow, in compound umbels; fruits small, dry, 2-seeded, narrowly winged.

CHEMISTRY Essential oil (fruits): **(+)-carvone** (70%), (+)-limonene (30–40%). Fruits of Indian chemotype (*A. sowa*): dillapiole (to 40%).

PHARMACOLOGY Dill fruits (and the oil) have known antispasmodic and bacteriostatic activity.

TOXICOLOGY Dill is a culinary herb and spice.

Anethum graveolens L. (Apiaceae); *aneth* (French); *Dill* (German); *aneto* (Italian); *eneldo* (Spanish)

Angelica archangelica
garden angelica • archangel

CLASSIFICATION TM: Europe. Comm.E+, ESCOP Suppl.

USES & PROPERTIES Roots (*Angelicae radix*) are used as appetite stimulant, spasmolytic and stomachic. Infusion or tinctures: 4.5 g of dry root/day; essential oil: 10–20 drops/day. Extracts are ingredients of commercial digestive medicines. The whole herb (*Angelicae herba*), fruits (*Angelicae fructus*) or essential oil (*Oleum angelicae*) are sometimes used.

ORIGIN Europe and Asia (now widely cultivated).

BOTANY Robust biennial herb; leaves compound, base sheathing; stems thick, hollow; flowers green, in globose umbels; fruits flat, winged, 2-seeded.

CHEMISTRY Essential oil rich in furanocoumarins (e.g. **angelicin**, imperatorin, xanthotoxin) and coumarins (e.g. osthole, osthenole, umbelliferone).

PHARMACOLOGY Digestive, antispasmodic and cholagogue activities have been demonstrated.

TOXICOLOGY Furanocoumarins are phototoxic. Side effects: skin irritation and allergic reactions.

NOTES In North America, *A. atropurpurea* is used.

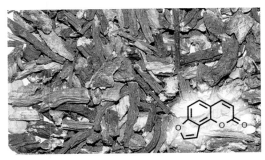

Angelica archangelica L. (Apiaceae); *archangélique* (French); *Engelwurz* (German); *archangelica* (Italian); *angélica* (Spanish)

Angelica sinensis
dang gui • Chinese angelica

CLASSIFICATION TM: Asia (China); Pharm.; WHO 2, HMPC.

USES & PROPERTIES Roots of *A. sinensis* (*Angelicae sinensis radix*) and *bai zhi* (*A. dahurica*) are traditionally used as tonics to treat anaemia, constipation, irregular menstruation, pain, chronic hepatitis and cirrhosis of the liver. The recommended daily dose is 4.5–9 g. Chinese angelica is second only to ginseng in importance as tonic.

ORIGIN Asia, East Asia.

BOTANY Perennial herb; stems purple; leaves large, pinnate; flowers small, white; fruits flat, prominently winged, 2-seeded.

CHEMISTRY The roots contain essential oil rich in alkylphthalides (the main compound is **ligustilide**); also non-volatile phenylpropanoids and coumarins.

PHARMACOLOGY Smooth muscle contraction, antihepatotoxic and analgesic activities.

TOXICOLOGY The roots are not toxic but treatment should be avoided during pregnancy (and by those using aspirin or warfarin).

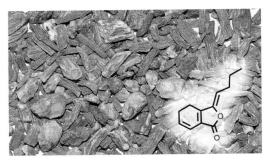

Angelica sinensis (Oliv.) Diels (Apiaceae); *angélique Chinoise*, *angélique de Chine* (French); *Chinesische Engelwurz* (German); *Kinesisk angelikarot*, *Kinakvanne* (Swedish)

Apium graveolens
celery

CLASSIFICATION TM: Europe, Asia.
USES & PROPERTIES The small dry fruits (*seeds*) (*Apii fructus*) are used as diuretic (kidney and bladder ailments), as adjuvant (rheumatism, arthritis), as stomachic and carminative (for nervous conditions) and in anti-inflammatory mixtures.
ORIGIN Africa, Asia, Europe.
BOTANY Erect, biennial herb; leaves large, pinnate; flowers in umbels; fruits small, dry, 2-seeded.
CHEMISTRY Essential oil (fruits) contains several aromatic compounds such as limonene and apiole. Of special interest are two minor compounds, **sedanolide** (1%) and 3-butylphthalide (1%).
PHARMACOLOGY Activity is ascribed to the essential oil compounds: 3-butylphthalide and 3-butyl-4,5-dihydrophthalide are anticonvulsant (mice, rats); sedanolide and other methylphthalides have sedative and spasmolytic activities.
TOXICOLOGY Fruits are edible. 3-butylphthalide: LD_{50} = 2450 mg/kg (rats, p.o.).
NOTES Leaves, petioles and roots are variously used as vegetables, culinary herb and spice.

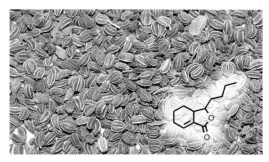

Apium graveolens L. (Apiaceae); *qin cài, sai kan choi* (Chinese); *célerei* (French); *Sellerie* (German); *sèdano* (Italian); *apío* (Spanish)

Aralia racemosa
American spikenard

CLASSIFICATION TM: North America; DS, homoeopathy.
USES & PROPERTIES Roots and root bark are used in traditional medicine as adaptogen, similar to ginseng. Also used for chronic cough and infections of the upper respiratory tract. The Asian *A. elata* and *A. mandshurica* have similar uses in Chinese Traditional Medicine.
ORIGIN United States and Canada.
BOTANY Woody shrub (to 2 m); leaves large, pinnately compound, deciduous; flowers small, whitish, in umbels; fruits fleshy, red to purple, not edible.
CHEMISTRY Poorly known; the related *A. elata* and *A. mandshurica* have triterpenoid saponins, of which **araloside A** is a main compound.
PHARMACOLOGY Limited information available; a root extract with a terpenoid: anticancer (cytotoxic) activity; araloside A: demonstrated anti-ulcer activity.
TOXICOLOGY Details not available.
NOTES A cultivated ornamental in the USA.

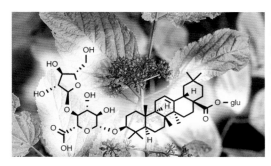

Aralia racemosa L. (Araliaceae); *Amerikanische Narde* (German)

Arctium lappa
burdock • greater burdock

CLASSIFICATION TM: Europe. Pharm., HMPC.
USES & PROPERTIES Dried roots (*Arctii/Bardanae radix*) are used (also in homoeopathy) for gastrointestinal ailments. Burdock leaf (*Bardanae folium*) is an ingredient of anti-inflammatory preparations (e.g. for rheumatism). Internal use: mostly cold infusions of root or leaf; external: root oil or powdered leaves.
ORIGIN Europe (introduced to Asia and North America); cultivated in Eastern Europe (and Japan, where the roots are eaten as vegetable).
BOTANY Robust biennial herb; leaves large, hairy below; flowers purple, in hairy and bristly heads.
CHEMISTRY Roots: lignans, essential oil (with e.g. benzaldehyde, acetaldehyde and pyrazines), inulin, triterpenes and especially various types of polyacetylenes (e.g. **arctinal**, lappaphens).
PHARMACOLOGY Anti-inflammatory, antimicrobial, hypoglycaemic and diuretic activities (ascribed to the lignans and polyacetylenes).
TOXICOLOGY No risks are known.
NOTES Alternatives: lesser burdock (*A. minus*) (in the UK) and woolly burdock (*A. tomentosum*).

Arctium lappa L. [= *A. majus* Bernh.] (Asteraceae); *gouteron, grateron* (French); *Große Klette* (German); *bardana* (Italian); *bardana* (Spanish)

Arctostaphylos uva-ursi
bearberry • uva-ursi

CLASSIFICATION TM: Europe. Pharm., Comm. E+, ESCOP 5, WHO 2, PhEur8, HMPC.
USES & PROPERTIES Dried leaves (*Uvae ursi folium*) are a traditional urinary antiseptic and ingredient of commercial preparations for kidney and bladder health. The daily dose is 10 g (equivalent of 400–700 mg of arbutin) or 2 g in 150 ml water, taken 3 or 4 times a day (maximum single dose of 3 g).
ORIGIN Arctic region (Europe, Asia, N. America).
BOTANY Woody shrub; branches trailing; bark reddish brown, smooth, flaking; flowers white, urn-shaped; fruit a bright red berry.
CHEMISTRY The leaves are chemically diverse (rich in tannins and acids) but **arbutin** and other hydroquinone derivatives are of special interest.
PHARMACOLOGY The free hydroquinone is the active compound which exhibits pronounced antimicrobial activities. It is not clear if the glucoside arbutin appears in the urine and where and how it is converted into the free aglycone.
TOXICOLOGY Do not use for prolonged periods: up to 7 days only; up to 5 treatments per year.

Arctostaphylos uva-ursi (L.) Spreng. (Ericaceae); *raisin d'ours, busserole officinale* (French); *Echte Bärentraube* (German); *uva ursina* (Italian); *gayuba del pays* (Spanish)

Areca catechu
areca nut

CLASSIFICATION Neurotoxin, stimulant (II). TM: Asia (taenicide – humans and animals).
USES & PROPERTIES Slices of betel nuts are habitually chewed by millions of people in Asia and Africa. It is mixed with betel vine leaves (*Piper betle*) and lime; tobacco, various spices, sweets and other flavourants may be added.
ORIGIN Cultigen of Asian origin; widely cultivated in tropical Asia, Africa and Madagascar.
BOTANY Single-stemmed palm tree (to 15 m or more); leaves pinnately dissected; fruit a fibrous 1-seeded drupe; seed globose, 20 mm in diameter.
CHEMISTRY Alkaloids (0.2–0.5%): **arecoline** (the main compound), with arecaidine, guvacoline, and guvacine. Lime is added to convert the alkaloids to their free bases. Condensed tannins (to 15%); bright red phlobatannins are formed.
PHARMACOLOGY Arecoline is a parasympathomimetic which acts on muscarinic receptors (and also nicotinic receptors, at high doses).
TOXICOLOGY Fatal dose: 8–10 g of seed; increased risk of oral cancer.

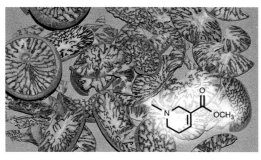

Areca catechu L. (Arecaceae); *aréquier* (French); *Betelnusspalme* (German); *avellana d'India* (Italian)

Argemone ochroleuca
prickly poppy • Mexican poppy

CLASSIFICATION Neuro- and cell toxin, mind-altering (II). TM: C and S America, India, Africa.
USES & PROPERTIES The yellow sap is widely used against skin ailments. Seeds may contaminate cereals such as wheat (with fatal results) and are an adulterant of mustard seed and mustard oil (resulting in epidemic dropsy). It is an addictive stimulant, marijuana substitute and aphrodisiac.
ORIGIN Central and South America; now a common weed in Africa, Europe and Asia.
BOTANY Annual herb (to 1 m); stems and leaves prickly, with bright yellow sap; flowers bright (*A. mexicana*) or pale yellow (*A. ochroleuca*); fruit a many-seeded capsule; seeds small, globose, black.
CHEMISTRY Isoquinoline alkaloids: berberine and protopine (plant); **sanguinarine** (seeds).
PHARMACOLOGY Berberine and sanguinarine are potentially carcinogenic. Berberine: moderately toxic; sanguinarine: toxic, linked to glaucoma.
TOXICOLOGY Sanguinarine: LD_{50} = ca. 18 mg/kg (mouse, i.p.); seed oil: at least 8.8 ml/kg required to produce toxicity symptoms in humans.

Argemone ochroleuca Sweet (Papaveraceae); *pavot épineux* (French); *Stachelmohn* (German); *pavero messicano* (Italian)

Aristolochia clematitis
birthwort

CLASSIFICATION   Cell toxin, mutagen (II). TM: Europe (no longer allowed).

USES & PROPERTIES   The whole herb (*Aristolochiae herba*) was traditionally used to induce labour, abortion and menstruation. The flowers resemble the human foetus in shape and orientation, hence the uses associated with childbirth. Externally it has been applied to sores and wounds. *Aristolochia* species (and *Asarum europaeum*, photo below) are no longer considered safe to ingest.

ORIGIN   Southern and central Europe.

BOTANY   Perennial herb with creeping rhizomes; stems erect; leaves heart-shaped; flowers yellow.

CHEMISTRY   The main compounds of interest are aristolochic acids I–IV.

PHARMACOLOGY Aristolochic acid I and others are nephrotoxins, potent mutagens and carcinogens (metabolically activated in the liver).

TOXICOLOGY   *Aristolochia* species are the known cause of several human fatalities; 70 cases of fibrous interstitial nephritis were recorded in Belgium.

Aristolochia clematitis L. (Aristolochiaceae); *sarrasine* (French); *Gewöhnliche Osterluzei* (German); *aristolochia* (Italian); *aristoloquia* (Spanish)

Armoracia rusticana
horseradish

CLASSIFICATION   Cell toxin (III). TM: Europe. Pharm., Comm.E+.

USES & PROPERTIES   The fresh root (*Armoraciae radix*) is traditionally ingested to treat bronchial and urinary tract infections (ca. 20 g/day). Preparations with 2% mustard oil are applied topically as counter-irritant for treating inflammation and rheumatism. Best known as a pungent condiment and source of the enzyme peroxidase.

ORIGIN   Uncertain (SE Europe or W Asia); a partly sterile cultigen, cultivated as a crop (root cuttings).

BOTANY   Stemless leafy perennial with a thick taproot; leaves oblong; flowers white.

CHEMISTRY   Glucosinolates (also known as mustard oil glycosides); sinigrin (the main compound) is enzymatically converted to the volatile, pungent and highly reactive **allyl isothiocyanate**.

PHARMACOLOGY   Isothiocyanates form covalent bonds with proteins: they are antimicrobial, cytotoxic, spasmolytic and skin irritant (hyperaemic).

TOXICOLOGY   Very high oral doses can be fatal; 3 g allyl isothiocyanate is a lethal dose in cattle.

Armoracia rusticana P. Gaertn., Mey. & Scherb. [= *Armoracia lapathifolia* Gilib.] (Brassicaceae); *grand raifort* (French); *Meerrettich* (German); *cren* (Italian); *rábano picante* (Spanish)

Arnica montana
arnica

CLASSIFICATION Cell toxin (II). TM: Europe;. Pharm., Comm.E+, ESCOP 4, WHO 3, HMPC.
USES & PROPERTIES Mainly the flower heads (*Arnicae flos*) are used (topically only!) for bruises, burns, diaper rashes, haematomas, sprains and sunburn. It can also be used as a mouthwash and gargle to treat inflammation of the mucous membranes. Infusions or tinctures (2 g dry herb in 100 ml) are used (diluted when used as mouthwash). Ointments contain 20–25% tincture.
ORIGIN Europe; the North American *A. chamissonis* and *A. fulgens* are now the main sources.
BOTANY Perennial herb; leaves oblong, hairy; flower heads large, deep yellow.
CHEMISTRY Sesquiterpenoids (0.2–0.5%): **helenalin** is the main compound.
PHARMACOLOGY Helenalin (when used topically): analgesic, anti-inflammatory, antiseptic, hyperaemic and wound-healing.
TOXICOLOGY Sesquiterpenes bind to proteins, causing allergic reactions; potentially lethal when ingested. Avoid contact with eyes or open wounds.

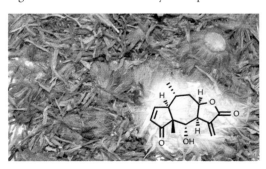

Arnica montana L. (Asteraceae); *arnica* (French); *Arnika, Bergwohlverleih* (German); *arnica* (Italian); *arnica* (Spanish)

Aronia melanocarpa
black chokeberry

CLASSIFICATION TM (DS): N America, Europe.
USES & PROPERTIES The small black berries (ripe or dried) or juices/concentrates became popular as dietary supplements and functional foods. Therapeutic value: reduction of oxidative stress (due to high levels of reactive oxygen free radicals) resulting from injury, operations, chemotherapy and chronic ailments (metabolic syndrome, diabetes).
ORIGIN North America (USA and Canada).
BOTANY Woody shrub (to 2 m); flowers white; fruit a small berry (ca. 1 g).
CHEMISTRY Polyphenols, especially proanthocyanidins; anthocyanins: **cyanidin 3-O-galactoside** (the main pigment, up to 14.8 mg/g); organic acids; vitamin C (ca. 50 mg per 100 g).
PHARMACOLOGY Health benefits are ascribed to the free radical scavenging and antioxidant activity of the proanthocyanidins, anthocyanidins and vitamin C. The antioxidant flavonoids can modulate several targets, mediating possible venotonic and anti-inflammatory effects.
TOXICOLOGY The berries are edible (non-toxic).

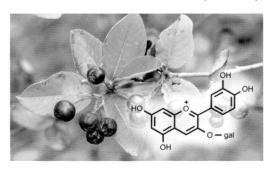

Aronia melanocarpa (Michx.) Elliot [= *Mespilus arbutifolia* L. var. *melanocarpa*] (Rosaceae); *hei guo xian lei hua qiu* (Chinese); *aronie à fruits noirs* (French); *Schwarze Apfelbeere* (German)

Atropa belladonna
deadly nightshade • belladonna

CLASSIFICATION Neurotoxin, hallucinogen (Ia); TM: Europe. Comm.E+, PhEur8. MM: (atropine).

USES & PROPERTIES Leaves (*Belladonnae folium*) and roots (*Belladonnae radix*) are famous for their traditional uses in treating pain and asthma. The isolated alkaloids (up to 2.2 mg per day) have many uses in modern medicine, as tranquillisers, spasmolytics and eye drops (to dilate the pupil of the eye, e.g. for eye diagnosis). Belladonna is infamous for its aphrodisiac and hallucinogenic uses.

ORIGIN Europe, Asia and North Africa.

BOTANY Perennial herb; leaves simple, lanceolate; flowers brownish-pink; fruit a shiny black berry.

CHEMISTRY Tropane alkaloids (1% in leaves, 2% in roots): **L-hyoscyamine** is the main compound in fresh leaves. It forms a racemic mixture called atropine when the leaves are dried.

PHARMACOLOGY Hyoscyamine: depressant and sedative at low doses; analgesic, hallucinogenic and potentially fatal at high doses.

TOXICOLOGY Atropine: the lethal oral dose is 10 mg in adults.

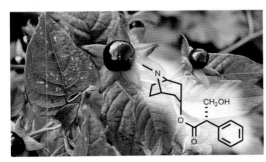

Atropa belladonna L. (Solanaceae); *belladonne, morelle furieuse* (French); *Tollkirsche* (German); *belladonna* (Italian); *belladonna* (Spanish)

Avena sativa
oats

CLASSIFICATION TM (DS): Europe. Comm.E+ (oat straw only), HMPC.

USES & PROPERTIES Aerial parts, harvested just before full flowers, are known as "oats green tops" (*Avenae herba recens*) and are believed to have sedative effects. Oats straw (*Avenae stramentum*) is added to bath water to treat skin ailments. Ripe oats fruits (*Avenae fructus*) are considered beneficial as a dietary supplement for relief of digestive ailments and general weakness. Oat bran (100 g per day) is believed to lower cholesterol levels.

ORIGIN Mediterranean region (southern Europe and North Africa) to Ethiopia. A major crop.

BOTANY Annual grass (to 1 m); spikelets pendulous, with persistent glumes.

CHEMISTRY Soluble silica and minerals; amino acids; B-vitamins; polysaccharides; triterpene saponins; **gramine** (an indole alkaloid).

PHARMACOLOGY The therapeutic benefits of oats straw are associated with silica and minerals. Gramine may have a sedative effect.

TOXICOLOGY Oats products are not toxic.

Avena sativa L. (Poaceae); *avoine* (French); *Hafer* (German); *avena* (Italian); *avena* (Spanish)

Azadirachta indica
neem tree • neem • nim

CLASSIFICATION TM: Asia (Ayurveda). WHO 3. Natural insecticide.
USES & PROPERTIES The bark, leaves, twigs and seeds are all used. The leaves are famous for their insecticidal properties; watery extracts are used by farmers as a natural and cheap insecticide. The seed oil, and infusions and extracts of bark and leaves are traditionally used to treat wounds, skin infections, stomach ailments, haemorrhoids, malaria and intestinal parasites. An ingredient of commercial skin care products (soaps, lotions).
ORIGIN South Asia (India, Sri Lanka and Burma); cultivated in Asia, Africa and Latin America.
BOTANY Evergreen tree (10 m or more); leaves compound, the leaflets asymmetrical, margins serrate; flowers small, white; fruit fleshy, 1-seeded.
CHEMISTRY Tetranortriterpenoids (limonoids) with **azadirachtin** as one of the main compounds.
PHARMACOLOGY Azadirachtin is an insect antifeedant, disrupting the metabolism of moth larvae.
TOXICOLOGY Seed oil: LD_{50} = 14 ml/kg (rat, p.o.); known to be toxic if taken internally.

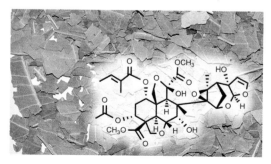

Azadirachta indica A. Juss. (Meliaceae); *margousier, neem* (French); *Nimbaum, Neembaum* (German); *nem* (Italian); *margosa* (Spanish)

Bacopa monnierii
brahmi • water hyssop

CLASSIFICATION TM: Asia (Ayurveda).
USES & PROPERTIES The herb or extracts are traditionally used as a nerve tonic and to improve learning. Daily dose: 300 mg of extract.
ORIGIN Asia (India); widely naturalised and cultivated (marshy areas), e.g. USA (Florida, Hawaii).
BOTANY Creeping perennial herb; leaves small, succulent; flowers white or pink. The name brahmi is also used for *gotu kola* (*Centella asiatica*): apparently a mistake dating from the 16th century.
CHEMISTRY Complex mixtures of triterpenoid saponins (called bacosides and bacopasides, e.g. **bacoside B**). Bacoside A is often studied (it is actually a mixture of several saponins, with bacoside A3 as one of the main constituents).
PHARMACOLOGY The saponins are believed to enhance cognitive function and have neuroprotective activity. They inhibit acetylcholinesterase and increase cerebral blood flow.
TOXICOLOGY Hardly any side effects (rarely nausea, diarrhoea and upset stomach). Not toxic: LD_{50} = 2400 mg/kg (rat, p.o., single dose).

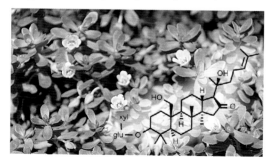

Bacopa monnieri (L.) Wettst. (Plantaginaceae); *jia ma chi xian* (Chinese); *bacopa de Monnier* (French); *brahmi* (Hindi); *Bacopa, Kleines Fettblatt, Wasser-Ysop* (German)

Brugmansia suaveolens
angel's trumpet

CLASSIFICATION Neurotoxin, hallucinogen, extremely hazardous (Ia). TM: South America.
USES & PROPERTIES Leaves, in the form of ointments applied to the skin, have been used as a traditional narcotic and hallucinogen by shamans.
ORIGIN Central and South America. A specific selection/cultigen called *Methysticodendron* has been grown by shamans using vegetative propagation to maintain the desired properties.
BOTANY Shrub or small tree; leaves large, velvety; flowers enormous, pendulous, attractive, fragrant.
CHEMISTRY Tropane alkaloids (to 0.3% in leaves): **scopolamine** (hyoscine), hyoscyamine, norhyoscine and tigliate esters of tropine.
PHARMACOLOGY The tropane alkaloids are mACh-R antagonists which work as parasympatholytics and cause euphoria and hallucinations (e.g. the feeling of being able to fly that may last for several hours). High doses lead to death by respiratory arrest. Mydriatic effects last up to six days.
TOXICOLOGY Cases of severe poisoning are quite common (see *Datura stramonium*).

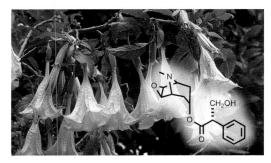

Brugmansia suaveolens (Willd.) Bercht. & C.Presl (Solanaceae); *stramoine odorante*, *Brugmansia* (French); *Engelstrompete* (German)

Bulbine frutescens
bulbine • burn jelly plant

CLASSIFICATION TM: Africa. AHP.
USES & PROPERTIES The colourless and slimy leaf parenchyma is traditionally applied to treat minor cuts, burns; wounds and itching (e.g. mosquito bites). The gel has been used as a famine food and has become popular as an ingredient of cosmetic and skin care products (comparable to *Aloe vera* gel).
ORIGIN Africa (South Africa); a popular garden succulent; grown on a small commercial scale.
BOTANY Perennial herb (ca. 0.2 m); leaves cylindrical, succulent; flowers small, yellow or orange; stamens hairy; fruit a small capsule.
CHEMISTRY The leaf contains polysaccharides and mucilages of as yet unknown composition. The roots contain anthraquinones (including chrysophanol and **knipholone**).
PHARMACOLOGY The soothing and anti-itching effects are ascribed to the demulcent and anti-inflammatory activities of the leaf gel.
TOXICOLOGY The gel is not toxic (edible).
NOTES Roots or stems of several *Bulbine* spp. are used in African traditional medicine.

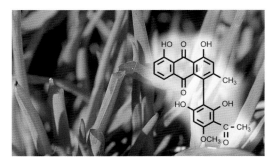

Bulbine frutescens Willd. [Xanthorrhoeaceae (formerly Asphodelaceae)]; *rankkopieva* (Afrikaans); *bulbine* (French); *Bulbine* (German)

Cajanus cajan
pigeon pea • pigeonpea

CLASSIFICATION   TM: Africa, Asia. AHP.
USES & PROPERTIES   Seeds are used in preparations to treat sickle cell anaemia. The roots, leaves, flowers and seeds have numerous medicinal uses. Infusions of the leaves: anaemia, diabetes, hepatitis, urinary infections and yellow fever. Flower infusions: dysentery, menstrual disorders. Immature fruits: kidney and liver ailments; fresh seeds are also used against urinary incontinence.
ORIGIN   Asia (probably India), but spread to Africa and the rest of Asia centuries ago.
BOTANY   Shrub (to 4 m); leaves trifoliolate; flowers yellow or multi-coloured; pods 2–9-seeded.
CHEMISTRY   Free amino acids in seeds: **phenylalanine** (5 mg/g); also hydroxybenzoic acid (21 mg/g). Leaves: prenylated stilbenes (longistylin A and C), flavonoids and triterpenoids.
PHARMACOLOGY   Activity against sickling (proven clinically) is ascribed to phenylalanine and hydroxybenzoic acid. Longistylin A is cytotoxic.
TOXICOLOGY   Seeds are edible (but contain trypsin and chymotrypsin inhibitors).

Cajanus cajan (L.) Millsp. (Fabaceae) *mu do* (Chinese); *pois cajun* (French); *arhar, tuur* (Hindi); *Straucherbse* (German); *caiano* (Italian); *gandul* (Spanish)

Calendula officinalis
marigold • pot marigold

CLASSIFICATION   TM: Europe. Pharm., Comm. E+, ESCOP 1, WHO 2, HMPC.
USES & PROPERTIES   Flower heads are used (*Calendulae flos, Calendulae flos sine calyce*), mainly for local applications to treat slow healing wounds, burns, eczema, dry skin, oral thrush and haemorrhoids. Preparations for external application contain 2–5 g of crude herb per 100 g. Infusions of 1–2 g of herb are taken 2–3 times per day for inflammation of the mouth and throat, indigestion (stimulation of bile production), gastric ulcers and menstrual disorders.
ORIGIN   Europe (widely cultivated).
BOTANY   Annual herb; leaves glandular, aromatic; flower heads large, shades of yellow and orange.
CHEMISTRY   Flavonoids (to 0.8%, e.g. isorhamnetin 3-O-glucoside); saponins (to 10%); triterpenes; essential oil (**α-cadinol**, α-ionone, β-ionone).
PHARMACOLOGY   Anti-inflammatory and antispasmodic effects have been demonstrated.
TOXICOLOGY   Edible (but use sparingly).
NOTES   Flowers add colour to herbal teas.

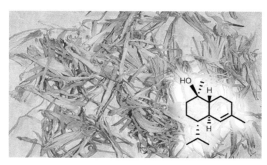

Calendula officinalis L. (Asteraceae); *souci des jardins* (French); *Ringelblume* (German); *calendola* (Italian); *caléndula* (Spanish)

Callilepis laureola
ox-eye daisy

CLASSIFICATION Cell toxin, highly hazardous (Ib). TM: Africa.
USES & PROPERTIES Weak infusions of the tuberous roots are used in traditional medicine to treat a wide range of ailments, including cough in adults. It is always expelled immediately after being drunk and is never taken as a enema. The Zulu name *impila* means "health". Traditionally, only weak infusions are used, and for adults only. However, fatal cases of poisoning regularly occur as a result of overdose and inappropriate use (e.g. when used to treat infants and young children).
ORIGIN Africa (eastern parts of South Africa).
BOTANY Perennial herb (to 0.5 m); leaves sparsely hairy; flower heads large with white ray florets.
CHEMISTRY The toxic compound is **atractyloside** (a kaurene glycoside).
PHARMACOLOGY Atractyloside inhibits ATP/ADP transport at the mitochondrial membrane, disrupting the energy supply within cells.
TOXICOLOGY A common cause of human fatalities. Atractyloside: LD_{50} = 431 mg/kg (rat, i.m.).

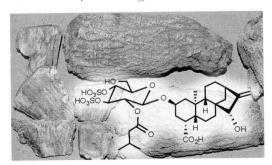

Callilepis laureola DC. (Asteraceae); *impila* (French); Impila (German); *impila, ihlmvu* (Zulu)

Camellia sinensis
tea • chai

CLASSIFICATION TM: Asia, East (China), Europe. Pharm., HMPC. DS: especially green tea.
USES & PROPERTIES Infusions of young leaves (*Theae folium*), usually accompanied by the unopened apical bud (*pekoe*) are used as a refreshing and stimulating beverage. Tea is used topically for weight loss and to treat skin ailments. Green tea (unfermented; heat-treated and rapidly dried) has become popular: antioxidant and diuretic, thought to be antimutagenic and anticarcinogenic.
ORIGIN Southern and eastern Asia. Cultivated in China since ancient times and later spread to India, Sri Lanka, Malaysia, Indonesia and Africa.
BOTANY Large shrub (pruned to 1.5 m for ease of harvesting); leaves glossy; flowers white.
CHEMISTRY Caffeine, phenolic acids and tannins.
PHARMACOLOGY Caffeine is a stimulant (it affects adenosine receptors). **Epigallocatechin gallate** is the most important catechin: antidiarrhoeal, antioxidant, diuretic and possible antimutagenic and anticholesterol activities).
TOXICOLOGY Safe, even in large amounts.

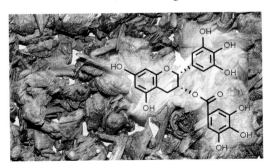

Camellia sinensis (L.) O.Kuntze (Theaceae); *chai* (Chinese); *théier* (French); *Teestrauch* (German); *tè* (Italian); *té* (Spanish)

Camptotheca acuminata
camptotheca • cancer tree

CLASSIFICATION Cell toxin, highly hazardous (Ib). TM: Asia (China). MM: clinical studies+.

USES & PROPERTIES The tree is used in traditional medicine in China to treat colds and psoriasis, as well as ailments of the stomach, spleen, liver and gall bladder. In modern medicine, pure alkaloid is administered by intravenous drip in cancer therapy (colorectal, ovarian and pancreatic cancers).

ORIGIN East Asia (China); cultivated as ornamental tree and as crop in India, Japan and the USA.

BOTANY Large tree (to 25 m); leaves large; flowers small, white, in rounded heads.

CHEMISTRY Pentacyclic quinoline alkaloids: **camptothecin** as main compound.

PHARMACOLOGY Camptothecin: cytostatic and antitumour activity (interrupts replication and transcription of nuclear DNA). Severe side effects (diarrhoea, haemorrhagic cystitis), so that several semisynthetic (and more soluble) analogues have been developed, e.g. topotecan, irinotecan.

TOXICOLOGY Camptothecin: LD_{50} = 50.1 mg/kg (mouse, p.o.).

Camptotheca acuminata Decne [Cornaceae (formerly Nyssaceae)]; *xi shu* (Chinese); *camptotheca* (French); *Glücksbaum* (German); *camptotheca* (Italian)

Canella winterana
winter bark

CLASSIFICATION TM: Central America.

USES & PROPERTIES The bark (called cinnamon bark, wild cinnamon or white cinnamon) is a traditional spice, aromatic tonic and antiseptic. It has been used to treat colds and poor circulation. It has been used as an insecticide.

ORIGIN Central America (Florida to Bahamas). Small-scale production continues.

BOTANY Evergreen shrub or tree (to 10 m); leaves bright green; flowers small; fruit fleshy, red.

CHEMISTRY The bark contains numerous drimane-type sesquiterpenes (**muzigadial** is one of the main compounds) and essential oil with pinene, eugenol and other monoterpenoids.

PHARMACOLOGY The sesquiterpenes have a peppery taste; they are dialdehydes which can bind to proteins, thus explaining the antiseptic and tonic effects. Muzigadial has antifeedant activity against the African army worm and also shows potent molluscicidal effects.

TOXICOLOGY The sesquiterpenoids are cytotoxic but more poisonous to insects than mammals.

Canella winterana L. (Canellaceae); *cannelle blanche* (French); *Weißer Zimtrindenbaum* (German); *curbana, macambo* (Spanish)

Cannabis sativa
marijuana • Indian hemp

CLASSIFICATION Mind-altering (III); TM: Asia (China, India).

USES & PROPERTIES Female flowers with leaves (*Cannabis indicae herba*) are known as marijuana and the resin of female flowers as hashish (both are smoked as popular recreational drugs). Seeds (*huo ma ren* in TCM) are a mild laxative. In recent years, the herb is used medicinally to treat the nausea of chemotherapy, the depression and lack of appetite of AIDS patients and glaucoma (to lower intra-ocular pressure).

ORIGIN Asia; cultivated in temperate regions, as oil and fibre crop or (often illegal) intoxicant.

BOTANY Erect annual (to 4 m); leaves palmate; flowers small, female and male on separate plants.

CHEMISTRY Phenolic terpenoids (cannabinoids): mainly **Δ^9-tetrahydrocannabinol** (THC).

PHARMACOLOGY THC: psychotropic (euphoria, relaxation, slurred speech); also analgesic, anti-emetic, bronchodilatory, spasmolytic, hypotensive.

TOXICOLOGY THC: LD_{50} = 43 mg/kg (mouse, i.v). Cannabis possession is illegal in most countries.

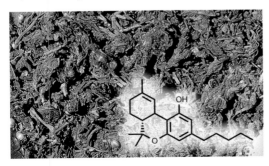

Cannabis sativa L. (Cannabaceae); *chanvre* (French); *Hanf* (German); *canapa indiana* (Italian); *cánamo* (Spanish)

Capsella bursa-pastoris
shepherd's purse • capsella

CLASSIFICATION TM: Europe. Comm.E+, HMPC.

USES & PROPERTIES The whole herb (*Bursae pastoris herba*) is traditionally used as tea (3–5 g, 10–15 g per day) to stop bleeding and to treat heavy periods (menorrhagia), diarrhoea and cystitis. It is applied to bleeding wounds or instilled in the nose to stop nosebleeds. In China the herb is used against eye diseases and dysentery. Commercial products (tinctures, tablets) are available.

ORIGIN Europe; now a cosmopolitan weed.

BOTANY Small annual or biennial herb (to 0.4 m); leaves in a rosette; flowers small, white; fruits shaped like a traditional shepherd's purse.

CHEMISTRY Diverse: amino acids, amines, flavonoids, monoterpenoids, glucosinolates, saponins.

PHARMACOLOGY The antihaemorrhagic activity is ascribed to a **peptide**; many other activities (but not linked to any single compound): anti-inflammatory, anti-ulcer, diuretic, urinary antiseptic and hypotensive effects.

TOXICOLOGY Avoid large doses.

peptide (unidentified)

Capsella bursa-pastoris L. (Brassicaceae); *bourse-à-pasteur* (French); *Hirtentäschel* (German); *borsa del pastore* (Italian); *bolsa de pastor* (Spanish)

Capsicum frutescens
chilli pepper • Tabasco pepper

CLASSIFICATION TM: South America, Europe. Pharm., Comm.E+ (external, capsaicin-rich species), ESCOP Suppl., HMPC, clinical studies+.
USES & PROPERTIES The whole dried fruit with seeds (*Capsici fructus acer*) and fruits of *C. anuum* (*Capsici fructus*) are used topically (in skin creams) for pain relief. Conditions treated include arthritis, rheumatism, neuralgia, lumbago, itching and spasms. It is taken against colic, dyspepsia and flatulence, or gargled to treat a sore throat.
ORIGIN Tropical America; a popular spice crop.
BOTANY Perennial herb (to 0.7 m); leaves glabrous; flowers white; fruit a berry, carried upright (drooping in *C. annuum*, an annual).
CHEMISTRY Pungent capsaicinoids: mainly **capsaicin**. The red pigments are carotenoids.
PHARMACOLOGY Capsaicin is a topical analgesic, carminative and counter-irritant. It causes a warm and painful sensation, followed by a long-lasting reversible desensitisation of nerve ends.
TOXICOLOGY Excessive use can be painful and may be dangerous (even lethal, in infants).

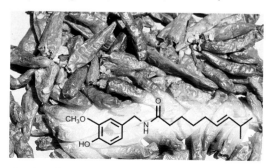

Capsicum frutescens L. (Solanaceae); *piment* (French); *Tabasco, Chili* (German); *peperoncino arbustivo* (Italian)

Carica papaya
papaya tree • paw paw

CLASSIFICATION TM: Central and South America. MM: papain (proteolytic enzyme).
USES & PROPERTIES The unripe fruits are a source of latex (papaya-latex, *papayotin*, *Caricae papayae succus*), from which papain is extracted. The crude or purified enzyme is included in digestive preparations and has been used (in doses of 1500 mg papain per day) to treat post-traumatic and post-operative swellings and oedemas. It is also used in wound treatment and injected to treat damaged intervertebral cartilage.
ORIGIN Central and South America. It is widely cultivated in the tropics.
BOTANY Small tree; leaves large, deeply lobed; flowers white; fruit a large, many-seeded berry.
CHEMISTRY Proteolytic enzymes (mainly **papain**) are extracted from the crude latex of unripe fruits.
PHARMACOLOGY Papain: digestive, wound-healing and anti-oedema activities have been demonstrated; it is a putative anthelmintic.
TOXICOLOGY The enzyme is not toxic but is best used in standardised preparations.

Carica papaya L. (Caricaceae); *papayer* (French); *Papaya, Melonenbaum* (German); *papaia* (Italian); *higo de mastuero* (Spanish)

Ceratonia siliqua
carob tree

CLASSIFICATION TM: Europe. DS: carob powder.
USES & PROPERTIES Powdered ripe fruits (carob powder) are used in specialised diets to treat diarrhoea and are thought to be of benefit in the management of coeliac (celiac) disorders. It is free from caffeine and theobromine and is therefore used as a chocolate substitute in the health food industry. The seed powder (locust bean powder) is a thickener, emulsifier and gelling agent in the food and pharmaceutical industries.
ORIGIN Mediterranean Europe (widely cultivated as a drought-tolerant street and fruit tree).
BOTANY Evergreen tree (to 15 m); leaves compound; flowers without petals (male and female on separate trees); fruit a many-seeded fleshy pod.
CHEMISTRY Polysaccharides (with **galactose** and mannose), sugars, crude fibre, tannins, proteins.
PHARMACOLOGY Antidiarrhoeal activity, probably due to tannins (they inactivate bacterial proteins). Soluble fibres are thought to prevent heart disease and lower serum cholesterol.
TOXICOLOGY The pulp of ripe fruits is edible.

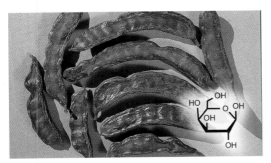

Ceratonia siliqua L. (Fabaceae); *caroubier* (French); *Johannisbrotbaum* (German); *carrubio* (Italian); *algarrobo* (Spanish)

Cerbera odollam
suicide tree • odollam tree

CLASSIFICATION Heart poison, extremely hazardous (Ia). TM: Pacific (Fiji).
USES & PROPERTIES Fruits of this tree (the bitter taste masked by sugar or spices) have the reputation of having killed more people (through suicide or murder) than any other plant poison. Hundreds of deaths have been reported from India. The related *C. manghas* of Madagascar is a traditional ordeal poison, responsible for thousands of deaths. Decoctions have been used as purgative medicine.
ORIGIN Asia (India, southeast Asia), Australia, Pacific region (sometimes grown in gardens).
BOTANY Evergreen shrub or small tree; flowers attractive, white; fruits large, fleshy, fibrous.
CHEMISTRY Cardiac glycosides (cardenolides): **cerberin** (=2'-acetylneriifoline) and several others.
PHARMACOLOGY Heart glycosides such as cerberin inhibit Na^+, K^+-ATPase in the heart muscle leading to cardiac arrest.
TOXICOLOGY Extremely toxic – death may follow within a few minutes of ingesting a concentrated decoction. Cerberin: LD_{50} = 0.147 mg/kg (cat, i.v.).

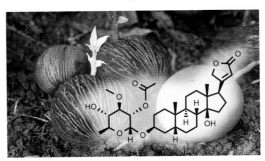

Cerbera odollam Gaertn. (Apocynaceae); *Zerberusbaum, See-Mango* (German)

Cetraria islandica
Iceland moss

CLASSIFICATION TM: Europe. Pharm., ESCOP, WHO 4, PhEur8, clinical studies+.

USES & PROPERTIES The whole herb (lichen thallus) is known as *Lichen islandicus* or *Cetrariae lichen*. Cold-water macerates: traditional bitter tonic (to reverse appetite loss); infusions (4–6 g of herb/day): cough, sore throat and gastroenteritis. Externally: applied to wounds to promote healing.

ORIGIN Arctic regions (wild-harvested).

BOTANY Small lichen; thallus branches up to 0.1 m high, brown above, greyish below.

CHEMISTRY Polysaccharides (mainly lichenin and isolichenin), up to 50% and bitter lichenolic acids (depsidones). The main acids are fumaroprotocetraric acid and **cetraric acid** (converted to other acids upon drying).

PHARMACOLOGY Antitussive and emollient (anti-irritant) activity, as well as immune-stimulant properties, are ascribed to the polysaccharides. The tonic properties and antibacterial activity are linked to the bitter-tasting lichenolic acids.

TOXICOLOGY Non-toxic at prescribed doses.

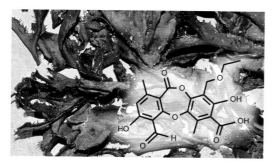

Cetraria islandica (L.) Ach. (Parmeliaceae); *lichen d'Islande* (French); *Isländisches Moos* (German); *lichene islandico* (Italian); *liquen islandico* (Spanish)

Chamaemelum nobile
Roman chamomile

CLASSIFICATION TM: Europe. Comm.E+, PhEur8, HMPC.

USES & PROPERTIES Dried flower heads (*Chamomillae romanae flos*) are traditionally used to treat stress-related dyspepsia (flatulence, nausea, vomiting and dysmenorrhoea). The essential oil (*Chamomillae romanae aetheroleum*) is used in cosmetics (e.g. shampoo) and in aromatherapy.

ORIGIN Europe. Widely cultivated.

BOTANY Perennial herb; leaves dissected; flower heads solid (hollow in German chamomile – *Matricaria chamomilla*) ray florets white.

CHEMISTRY Essential oil: bright blue azulenes (artefacts due to steam distillation); also sesquiterpenoids (mainly **α-bisabolene**, nobilin), flavones, organic acids and polyacetylenes.

PHARMACOLOGY The herb has demonstrated sedative, antispasmodic and anti-inflammatory activities. The volatile oil is antimicrobial.

TOXICOLOGY Large doses are emetic. Allergic and anaphylactic reactions are ascribed to the sesquiterpene lactones.

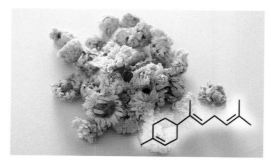

Chamaemelum nobile (L.) All. [= *Anthemis nobilis* L.] (Asteraceae); *camomille romaine* (French); *Römische Kamille* (German); *camomilla romana* (Italian); *manzanilla romana* (Spanish)

Chelidonium majus
greater celandine

CLASSIFICATION Cell toxin (II). TM: Europe. Comm.E+, ESCOP, WHO 5, PhEur8, HMPC.

USES & PROPERTIES The whole herb (*Chelidonii herba*) is traditionally used to treat spasms of the gastrointestinal tract and bile duct. The daily dose is 2–5 g (taken as infusion of 0.5–1 g, three times per day). It is believed to act as cholagogue (to stimulate bile flow) in treating hepatitis, jaundice and gall stones. There are many topical uses to treat warts, ringworm, eczema and eye complaints.

ORIGIN Europe, Asia and North Africa.

BOTANY Perennial herb with bright orange sap; leaves deeply lobed; flowers yellow.

CHEMISTRY Alkaloids: protopine, protoberberine and benzophenanthridine alkaloids: **chelidonine**, berberine, sanguinarine and others.

PHARMACOLOGY The alkaloids are DNA intercalculating and have antimicrobial, antispasmodic, analgesic and cholagogue (choleretic) activities.

TOXICOLOGY It is difficult to regulate the dose (overdosing causes side effects such as stomach cramps, dizziness). Possible liver toxicity.

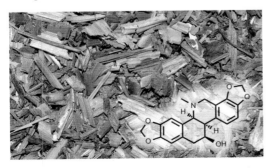

Chelidonium majus L. (Papaveraceae); *chélidoine, grande-éclaire* (French); *Schöllkraut* (German) *celidonia, cinerognola* (Italian); *celidonia* (Spanish)

Chenopodium ambrosioides
wormseed goosefoot

CLASSIFICATION Cell toxin; oil is highly hazardous (Ia). TM: South America, Europe. Pharm.

USES & PROPERTIES The essential oil (*Chenopodii aetheroleum*) is traditionally used as anthelmintic medicine in domestic animals and humans. The human dose is 1 g oil in castor oil (2.5% solution). Jesuit tea (*Tinctura botryos mexicanae*), made from the dried herb, was once used for abortion.

ORIGIN South and Central America (Mexico). Naturalised in many parts of the world.

BOTANY Erect perennial herb (to 1 m); leaves toothed; flowers minute.

CHEMISTRY Essential oil: **ascaridol** is the main compound.

PHARMACOLOGY Ascaridol has anthelmintic activity; it kills ascaris (maw worms) and hook worms in humans (and trematodes in animals).

TOXICOLOGY The essential oil is very toxic and may cause spasm and coma if the dose is not carefully controlled (maximum of 1 g per day). It is nowadays rarely used. Ascaridol: LD_{50} = 157 mg/kg (mouse, p.o.); lethal dose (rabbit): 0.6 ml/kg.

Chenopodium ambrosioides L. var. **anthelminticum** (L.) A. Gray (Amaranthaceae); *chénopode anthelmintique* (French); *Wurmtreibender Gänsefuß* (German)

Chondrodendron tomentosum
curare vine

CLASSIFICATION Neurotoxin, highly hazardous when injected (Ia); MM: muscle relaxant.

USES & PROPERTIES Pure alkaloid has been used as skeletal muscle relaxant in surgery (anaesthesia) but safer alternatives are nowadays available. The alkaloids are components of blow dart poisons used by South American Indians to ensure that prey animals (birds, monkeys) fall to the ground (because of the muscle-relaxant effect); the meat is safe to consume because the alkaloids are only poisonous when injected into the bloodstream.

ORIGIN Central and South America.

BOTANY Woody climber; leaves silver below; flowers small; fruit fleshy; seed halfmoon-shaped.

CHEMISTRY Dimeric isoquinoline alkaloids: **D-tubocurarine** (the major compound) and others.

PHARMACOLOGY Tubocurarine is a muscle-relaxant (binds to nACh-R as an antagonist). Injection causes paralysis, coma and death.

TOXICOLOGY Tubocurarine: LD_{50} = 0.23–0.70 mg/kg (various animals, i.v.); the lethal dose in rabbits (s.c.) is 3–5 mg.

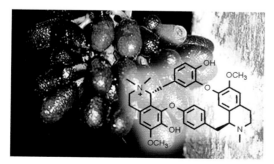

Chondrodendron tomentosum Ruiz & Pav. (Menispermaceae); *vigne sauvage, curare* (French); *Behaarter Knorpelbaum, Pareira* (German); *pareira brava* (Spanish)

Chondrus crispus
Irish moss • carragheen

CLASSIFICATION TM: Europe. PhEur8.

USES & PROPERTIES The dried whole algae (or isolated polysaccharides) are traditionally used to treat inflammation of the upper respiratory tract and stomach (especially cough and bronchitis). Used as bulk-forming laxatives and also as non-digestible adjuncts in weight loss diets.

ORIGIN North Atlantic region (Europe, America).

BOTANY A small seaweed (red algae) that grows on rocks along the intertidal zone.

CHEMISTRY The main compounds of interest are polysaccharides known as **carrageenans**. They are galactans (polymers of sulfated galactose). Also present are amino acids, proteins and iodine and bromine salts.

PHARMACOLOGY Polysaccharides are anti-irritant (demulcent), also thought to be mildly expectorant and laxative.

TOXICOLOGY Non-toxic but high doses of iodine may be harmful.

NOTES A source of polysaccharides used in the food, pharmaceutical and cosmetics industries.

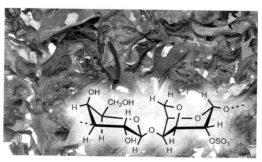

Chondrus crispus (L.) Stackh. (Gigartinaceae); *mousse d'Irlande* (French); *Knorpeltang, Irisch Moos* (German)

Chrysanthemum cinerariifolium
pyrethrum • Dalmation chrysanthemum

CLASSIFICATION Neurotoxin (II). Natural insecticide: Europe.
USES & PROPERTIES Powdered dried flower heads were once sold as lice remedies (as Persian powder: brand name Zacherlin). Flower heads and/or seed extracts have become commercial sources of natural, biodegradable insecticide.
ORIGIN Eastern Europe (Balkans). It is widely cultivated (Kenya, Tanzania, Ecuador, Tasmania).
BOTANY Perennial herb (to 0.8 m); leaves compound, deeply lobed, silvery; flower heads white.
CHEMISTRY The toxins are natural pyrethrins, with **pyrethrin I** and II as the main compounds. Synthetic pyrethroids are also used.
PHARMACOLOGY Pyrethrins are much more toxic to insects than to humans, with a rapid knockout effect. Piperonyl butoxide is added to synergistically enhance the toxicity to insects.
TOXICOLOGY Pyrethrin I: LD_{50} = 1.2 g/kg body weight (rat, p.o.). Synthetic pyrethroids are more toxic to humans than the natural ones. Both can cause severe allergic reactions in sensitive persons.

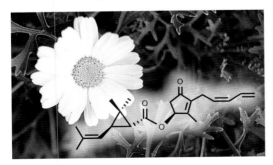

Chrysanthemum cinerariifolium (Trevir.) Vis. or **Tanacetum cinerariifolium** (Trevir.) Sch. Bip. (Asteraceae); *pyrèthre* (French); *Dalmatiner Insektenblume, Pyrethrum* (German)

Chrysanthemum ×morifolium
chrysanthemum

CLASSIFICATION TM: Asia (China). Pharm.
USES & PROPERTIES An infusion of a few dried flower heads (*Chrysanthemi flos*) is a popular general tonic and anti-inflammatory in Chinese medicine. It is widely used as an ingredient of medicinal teas to treat fevers, high blood pressure, infections and sore eyes. Powdered herb or poultices are applied to acne, sores, boils and skin infections.
ORIGIN Eastern Asia. A cultivar with small white flower heads (used for medicine) is commercially cultivated in China. Many other cultivars are grown all over the world as cut flowers.
BOTANY Perennial herb; leaves deeply lobed, aromatic; flower heads variable, in many colours.
CHEMISTRY Sesquiterpene alcohols: e.g. **chrysanthediol A**; triterpenes: e.g. helianol.
PHARMACOLOGY Chrysanthediol A and other terpenes are anti-inflammatory (and cytotoxic against human cancer cells). The herb is antimicrobial and appears to lower blood pressure.
TOXICOLOGY The flower heads are edible.

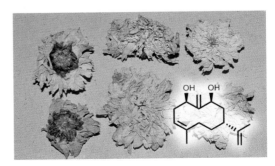

Chrysanthemum ×morifolium Ramat. (Asteraceae); *ju hua* (Chinese); *chrysanthème des fleuristes* (French); *Garten-Chrysantheme* (German); *chrysantemo* (Italian)

Cichorium intybus
chicory

CLASSIFICATION TM: Asia and Europe. Pharm., Comm.E+, HMPC.

USES & PROPERTIES The whole herb (*Chicorii herba*) is used as bitter tonic and "blood purifier" to treat dyspepsia and a loss of appetite. It is a traditional cholagogue, choleretic, carminative and diuretic in both Ayurvedic and European medicine. Chicory syrup is used as a tonic (infants) and cleansing medicine in cases of gout and rheumatism. Roots (*Chicorii radix*) are used to produce chicory (a coffee substitute and additive).

ORIGIN Europe and Asia (the wild form is naturalised in many parts of the world); commercial cultivars are grown for their root (source of chicory) or their leaves (vegetable and salad).

BOTANY Erect perennial herb (to 1 m); leaves large, dentate; flower heads sessile, pale blue.

CHEMISTRY Sesquiterpene lactones: **lactucin**.

PHARMACOLOGY Lactucin and other sesquiterpenes are responsible for the bitter taste. The herb has mild choleretic and diuretic activities.

TOXICOLOGY Non-toxic.

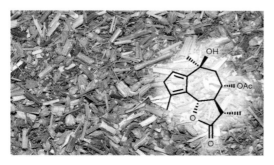

Cichorium intybus L. (Asteraceae); *chicorée sauvage* (French); *Wegwarte* (German); *cicoria* (Italian)

Cicuta virosa
cowbane • water hemlock

CLASSIFICATION Cell toxin, highly hazardous (Ib). TM: Europe. Pharm.

USES & PROPERTIES The herb (*Cicutae herba*) has been used in Europe as an antispasmodic and to alleviate the symptoms of rheumatism.

ORIGIN Northern Europe, northern Asia and North America (found in wet places).

BOTANY Perennial herb with thick rhizomes; Leaves compound; flowers small, white, in umbels; fruit a small (2 mm long) schizocarp.

CHEMISTRY Polyacetylenes (0.2% in fresh material, 3.5% in dried rhizomes): **cicutoxin** and cicutol are the main compounds.

PHARMACOLOGY Analgesic and antispasmodic activities; antileucaemic effects. The polyacetylenes are highly reactive and form covalent bonds with macromolecules in the cell, causing CNS symptoms and cell death.

TOXICOLOGY The LD_{50} for cicutoxin = 9.2 mg/kg (mouse, i.p.).

NOTES Related plants (*Cicuta maculata*, *Aethusa cynapium* and *Oenanthe crocata*) are also toxic.

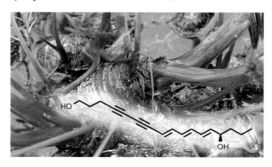

Cicuta virosa L. (Apiaceae); *cigué aquatique* (French); *Giftiger Wasserschierling* (German); *cicuta aquatica* (Italian)

Cimicifuga racemosa
black cohosh • black snakeroot

CLASSIFICATION TM: North America. Comm. E+, WHO 2, PhEur8, HMPC, clinical studies+.

USES & PROPERTIES Dried rhizomes and roots (*Cimicifugae racemosae rhizoma*) are traditionally used against premenstrual and menopausal disorders (0.5–1 g, up to three times per day). It is included in tonics and cough mixtures and used to treat chorea, rheumatism, dizziness and tinnitus.

ORIGIN North America (Canada and northeastern USA). Raw material is mostly wild-harvested.

BOTANY Perennial herb; leaves pinnately compound; flowers small, white, in elongated clusters.

CHEMISTRY Tetracyclic triterpenoids glycosides: **actaein** and cimicifugoside (and their aglycones, cimigenol and acetylacteol). Also formononetin (an isoflavonoid), gallotannins and organic acids.

PHARMACOLOGY Actaein is considered to be spasmolytic, vasodilatory and hypotensive. The drug has an oestrogen-like action.

TOXICOLOGY More than 5 g can cause toxic effects. Continuous use should be avoided.

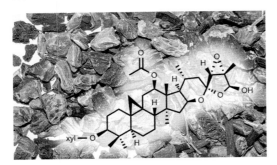

Cimicifuga racemosa (L.) Nutt. or ***Actaea racemosa*** L. (Ranunculaceae); *actée à grappet* (French); *Amerikanische Traubensilberkerze* (German); *cimicifuga* (Italian, Spanish)

Cinchona pubescens
Peruvian bark tree • red cinchona

CLASSIFICATION Cell toxin (II). TM: South America, Europe. Comm.E+ (tonic), PhEur8. MM: alkaloids (quinine, quinidine).

USES & PROPERTIES Bark (*Cinchonae cortex*) is traditionally used as bitter tonic (appetite stimulant) and treatment against malaria. Pure alkaloids (and synthetic derivatives) are used as antimalarial and anti-arrhythmic drugs.

ORIGIN South America (Colombia, Ecuador, Peru). Cultivation: India, Indonesia and Africa.

BOTANY Evergreen tree; bark reddish; leaves simple, large. Species such as *C. officinalis* (yellow bark) are also used for alkaloid extraction.

CHEMISTRY Quinoline alkaloids (5–15%): **quinine** is the main compound (also quinidine).

PHARMACOLOGY Quinine and synthetic derivatives disrupt the metabolism of *Plasmodium* parasites but many strains have become resistant. Quinidine inhibits Na^+-channels and is used as an anti-arrhythmic drug.

TOXICOLOGY Quinine has a relatively low toxicity and is used in tonic water (ca. 67 mg per litre).

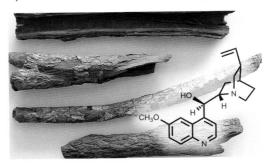

Cinchona pubescens Vahl [= *Cinchona succirubra* Pav. ex Klotsch] (Rubiaceae); *quina, quinquina* (French); *Roter Chinarindenbaum* (German); *china rossa* (Italian)

Cinnamomum camphora
camphor tree

CLASSIFICATION TM: East Asia (China), Europe. Pharm., Comm.E+.

USES & PROPERTIES Wood is distilled to obtain a camphor fraction (gum camphor) and a white essential oil rich in cineole (*Cinnamomi camphorae aetheroleum*). Both are used as circulatory and respiratory stimulants, as counter-irritants and ingredients of ointment (for rheumatism and congestion). It is used for relief of colds, fever, influenza, inflammation, pneumonia and diarrhoea.

ORIGIN China, Japan and Taiwan. Widely introduced as a popular ornamental tree.

BOTANY Large evergreen tree (to 50 m); leaves glossy, indistinctly trinerved; fruit a black drupe.

CHEMISTRY Essential oil, with **camphor** and 1,8-cineole (= eucalyptol) as main compounds.

PHARMACOLOGY Camphor and 1,8-cineole have (synergistic) antiseptic, analeptic, carminative, counter-irritant, spasmolytic and stimulant effects. Stimulation of cold receptors in the nasal passages gives a cooling effect.

TOXICOLOGY Camphor is toxic in high doses.

Cinnamomum camphora (L.) J. Presl (Lauraceae); *camphrier du Japon* (French); *Kampferbaum* (German); *camfora* (Italian); *alcanfor* (Spanish)

Cinnamomum verum
cinnamon bark tree • Ceylon cinnamon

CLASSIFICATION TM: Asia (India and Sri Lanka), Europe. Comm.E+, WHO 1, PhEur8, HMPC.

USES & PROPERTIES The inner bark of branches and coppice shoots (cinnamon; *Cinnamomi ceylanici cortex*) is used as appetite stimulant and to treat indigestion, dyspeptic complaints and other conditions (e.g. inflammation, nausea, loss of appetite, diarrhoea). The daily dose is 2–4 g (bark) or 0.05–0.2 g (essential oil).

ORIGIN Asia (Sri Lanka and parts of India); trees are cultivated in many tropical regions.

BOTANY Evergreen medium-sized tree; leaves glossy, trinerved; fruit an oblong, black drupe.

CHEMISTRY The essential oil is dominated by **cinnamaldehyde** (65–80%), with smaller amounts of eugenol and others. Bark is rich in procyanidins.

PHARMACOLOGY Cinnamaldehyde has antispasmodic and antimicrobial activity; the bark and tannins are astringent and the procyanidins contribute to antioxidant effects.

TOXICOLOGY Non-toxic at low doses and widely used as a spice (true cinnamon).

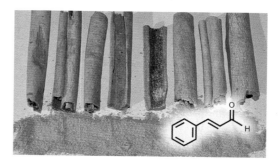

Cinnamomum verum J. Presl [= *C. zeylanicum* Nees] (Lauraceae); *canellier, canelle de Ceylan* (French); *Ceylon-Zimtbaum* (German); *cannella* (Italian); *canelo de Ceilán* (Spanish)

Citrullus colocynthis
colocynth • bitter apple

CLASSIFICATION Cell toxin (Ib). TM: Africa, Asia.

USES & PROPERTIES Fruits are no longer used in traditional medicine as purgatives (too toxic, side effects). Formerly used as insecticide and for rodent control. Seeds of wild watermelon (*Citrullus lanatus*) yield an edible oil, used in cosmetics.

ORIGIN West Africa (naturalised in Arabia, the Mediterranean region, Australia and India). Cultivated since Assyrian times.

BOTANY Annual trailing herb; leaves hairy; flowers yellow; fruit large, globose, many-seeded (resembling a small watermelon).

CHEMISTRY Cucurbitacins (bitter-tasting triterpenoids): **Cucurbitacins B**, E and J, together with their glycosides, are the main toxic compounds.

PHARMACOLOGY Cucurbitacins are cytotoxic; they have purgative, analgesic and antitumour activities but are too poisonous to be used medicinally.

TOXICOLOGY The lethal dose in humans is 3 g of the fruit. Exposure to skin may cause blisters.

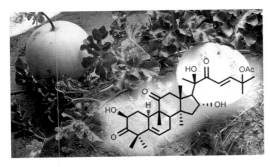

Citrullus colocynthis (L.) Schrad. (Cucurbitaceae); *coloquinthe* (French); *Koloquinte, Bittermelone* (German); *coloquintide* (Italian)

Citrus aurantium
bitter orange • Seville orange

CLASSIFICATION TM: Asia (China), Europe. Pharm., Comm.E+, PhEur8, HMPC.

USES & PROPERTIES The glandular outer fruit peel (*Aurantii pericarpium*), or essential oil extracted from it, are traditionally used as aromatic bitter tonic to stimulate appetite and to treat indigestion, flatulence and bloating. Unripe fruits (*zhi shi*) are used in Chinese medicine for relief of bloating. Ripe fruits are used to make marmalade.

ORIGIN Southern Asia. Cultivated as a fruit tree in most warm regions of the world.

BOTANY Evergreen tree (to 10 m); leaves glabrous, aromatic; flowers white; fruits depressedly ovate. The form used for bergamot oil (subsp. *bergamia*) is a smaller tree (to 5 m).

CHEMISTRY Bitter flavone glycosides (**naringenin**, neohesperidin); essential oil (limonene, linalool, terpineol); bitter triterpenes (limonin).

PHARMACOLOGY Bitter substances stimulate appetite; essential oil acts as aromatic and stomachic.

TOXICOLOGY Small amounts of phototoxic coumarins in the oil may cause photosensitisation.

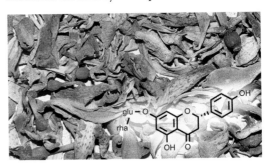

Citrus aurantium L. subsp. ***aurantium*** (Rutaceae); *orange amère* (French); *Pomeranze, Bitterorange* (German); *arancio amaro* (Italian); *naranjo amargo* (Spanish)

Coffea arabica
coffee tree • Arabian coffee

CLASSIFICATION Stimulant (III). TM: Africa, Europe. Comm.E+ (coffee charcoal). MM: caffeine.
USES & PROPERTIES Roasted seeds (*Coffeae semen*) are used medicinally as stimulant and diuretic. Caffeine is an ingredient of preparations used to treat fever, pain and influenza (and modern energy drinks). Coffee carbon (*Coffeae tostae carbo*) is taken orally (9 g per day) to treat diarrhoea and mild inflammation of the mouth and throat.
ORIGIN Africa (Ethiopia); cultivated commercially in many parts of the world.
BOTANY Evergreen shrub or small tree; leaves opposite; flowers fragrant; fruit a 2-seeded drupe.
CHEMISTRY Purine alkaloids: mainly **caffeine** (1–2% in ripe seeds, or 150 mg per cup).
PHARMACOLOGY Caffeine is a central stimulant (it inhibits adenosine receptors and cAMP phosphodiesterase) and has a positive inotropic action. Coffee enhances gastric secretions and gut motility.
TOXICOLOGY Caffeine is addictive. Excessive amounts may lead to high blood pressure, palpitations, nervousness, insomnia and indigestion.

Coffea arabica L. (Rubiaceae); *caféier d'Arabie* (French); *Kaffeestrauch* (German); *caffè* (Italian); *cafeto* (Spanish)

Cola acuminata
cola nut tree

CLASSIFICATION Stimulant. TM: Africa, Europe. Pharm., Comm.E+, PhEur8, HMPC.
USES & PROPERTIES The ripe dried seeds, with the seed coat removed (*Colae semen*) are traditionally used to treat mental and physical fatigue. Their astringency makes them useful to treat diarrhoea, wounds and inflammation. Cola nuts were formerly an ingredient of cola drinks and are still used in energy drinks.
ORIGIN West Africa (Nigeria, Sierra Leone to Gabon); cultivated in tropical Asia and America.
BOTANY Evergreen tree (to 15 m); leaves oblong; flowers yellow; fruit a large, multi-seeded follicle.
CHEMISTRY Purine alkaloids: caffeine (1.5–3%) and **theobromine**; phenolic compounds (4–6%).
PHARMACOLOGY The purine alkaloids have a stimulant activity (on the heart and central nervous system, see *Coffea*). The nuts have positive chronotropic and weak diuretic effects in humans.
TOXICOLOGY Non-toxic. Excessive amounts may be harmful to those with ulcers, heart disorders or hypertension.

Cola acuminata (Pal.) Schott & Endl. [Malvaceae (formerly Sterculiaceae)]; *colatier* (French); *Kolabaum* (German); *cola* (Italian); *cola* (Spanish)

Colchicum autumnale
autumn crocus • meadow saffron

CLASSIFICATION Cell toxin, extremely hazardous (Ia). TM: Europe. Comm.E+. MM: pure alkaloid.

USES & PROPERTIES Cut and dried corms (*Colchici tuber*), seeds (*Colchici semen*) or fresh flowers are used. Extracted alkaloids, in carefully controlled doses, are used in modern medicine to treat gout and familial Mediterranean fever. The maximum daily amount is 8 mg (administered orally, 0.5–1.5 mg every 2–3 hours). The plant and alkaloids are very toxic and not suitable for self-medication.

ORIGIN Europe and North America. The species and relatives are grown as garden plants.

BOTANY Deciduous bulbous plant; leaves in summer; flowers pink or purple, emerging in autumn.

CHEMISTRY Phenethylisoquinoline alkaloids: **colchicine** is the main compound.

PHARMACOLOGY Colchicine: anti-inflammatory and painkiller (prevents macrophages from reaching inflamed joints). The alkaloid is used as a spindle poison when studying chromosomes.

TOXICOLOGY Colchicine: the toxic dose is 10 mg; lethal dose 40 mg. All *Colchicum* species are toxic.

Commiphora myrrha
myrrh tree • African myrrh

CLASSIFICATION TM: Africa, Europe, Asia. Pharm., Comm.E+, ESCOP 6, PhEur8.

USES & PROPERTIES The oleoresin gum that exudes naturally from the bark (myrrh; *Myrrha*) has many uses in traditional medicine – mainly topically against mouth and throat infections, to promote wound healing and for treating various skin conditions (as antibiotic treatment and to stop bleeding and swelling). Myrrh tincture is dabbed onto wounds or 60 drops in a glass of warm water can be used as gargle or mouth rinse.

ORIGIN Northeastern Africa (Kenya, Ethiopia and Somalia). The resin is wild-harvested.

BOTANY Small deciduous tree (to 3 m); stems somewhat thorny; leaves small, on short shoots; flowers pink and yellow; fruits small, oval capsules.

CHEMISTRY Myrrh contains oleoresins, essential oil and polysaccharides. The essential oil (3–6%) contains about 50% **furanoeudesma-1,3-diene**.

PHARMACOLOGY Astringent, antiseptic, anti-inflammatory and antipyretic.

TOXICOLOGY Myrrh is non-toxic in small doses.

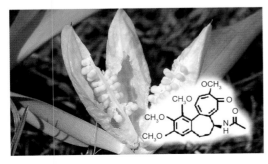

Colchicum autumnale L. (Colchicaceae); *colchique d'automne* (French); *Herbstzeitlose* (German); *crocus autumnale, colchico* (Italian); *cólquico* (Spanish)

Commiphora myrrha (Nees) Engl. [= *C. molmol* Engl.] (Burseraceae); *myrrhe* (French); *Myrrhe* (German); *mirra* (Italian); *mirra* (Spanish)

Conium maculatum
poison hemlock

CLASSIFICATION Neurotoxin, extremely hazardous (Ia). TM: Europe.
USES & PROPERTIES The herb or seeds have been used in traditional medicine as sedative, antispasmodic, antaphrodisiac and anti-ulcer treatment. It has also been used since ancient times for murder, suicide and execution. It was the main ingredient of the poison that allegedly killed Socrates.
ORIGIN Europe. Naturalised in Africa and Asia. Hemlock grows as a weed in disturbed places.
BOTANY Erect biennial herb (to 1.5 m); stems purple-spotted; leaves compound; flowers white. After handling the plant or its alkaloids, a characteristic mousy smell is left on the hands.
CHEMISTRY Piperidine alkaloids (up to 3.5% in fruits): mainly **coniine** and γ-coniceine, with several minor alkaloids. Toxic polyacetylenes in roots.
PHARMACOLOGY Coniine: sedative, antispasmodic, analgesic and extremely poisonous.
TOXICOLOGY Coniine: LD_{50} (mouse) = 19 mg/kg (i.v.), 100 mg/kg (p.o.). The lethal oral dose of coniine in humans is 0.5–1 g.

Conium maculatum L. (Apiaceae); *cigué tachée* (French); *Gefleckter Schierling* (German); *cicuta maggiore* (Italian)

Convallaria majalis
lily-of-the-valley

CLASSIFICATION Heart poison, highly hazardous (Ia). TM: Europe. Pharm., Comm.E+.
USES & PROPERTIES The whole herb (*Convallariae herba*), harvested during flowering, is traditionally used as a heart stimulant to treat the symptoms of mild cardiac insufficiency.
ORIGIN Europe and northeast Asia (naturalised in North America). Cultivated as ornamental plant. Raw material is harvested in eastern Europe.
BOTANY Small clump-forming perennial; leaves in a single pair; flowers white, bell-shaped; fruit a few-seeded berry, orange-red when ripe.
CHEMISTRY Several cardiac glycosides (0.1–0.5% of dry weight) and saponins. The main compound is **convallatoxin** (40% of total heart glycosides).
PHARMACOLOGY Convallatoxin is similar to other heart glycosides which inhibits Na^+, K^+-ATPase. It strengthens the contraction of the heart muscle and decreases the internal heart pressure.
TOXICOLOGY Convallatoxin is very poisonous: lethal dose = 0.5 g (humans, p.o.). Accidental poisoning (mistaken for *Allium ursinum*) is rarely fatal.

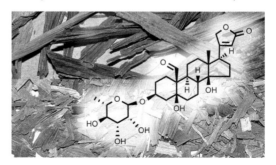

Convallaria majalis L. [Asparagaceae (formerly Convallariaceae)]; *muguet* (French); *Maiglöckchen* (German); *mughetto* (Italian); *lirio de los valles* (Spanish)

Coptis chinensis
Chinese goldthread

CLASSIFICATION TM: Asia (China). Pharm., WHO 1.

USES & PROPERTIES The rhizomes (*Coptidis rhizomae*), yellow or orange inside, are traditionally used in Chinese medicine to treat bacterial diarrhoea, gastroenteritis, inflammation, ulcers, sores and conjunctivitis. The daily dose of crude product is 1.5–6 g. Other species are used as bitter tonic and stomachics, and to treat oral infections.

ORIGIN Eastern Asia (China, also grown there).

BOTANY Perennial herb; leaves compound, on long stalks; flowers white or pale pink. Several other species are used in India (*C. teeta*), Japan (*C. japonica*) and North America (*C. trifolia*).

CHEMISTRY Isoquinoline (protoberberine) alkaloids: **berberine** is the main compound (5–7%).

PHARMACOLOGY The DNA intercalating berberine has documented antimicrobial activity against a wide range of organisms (see *Berberis vulgaris*).

TOXICOLOGY Berberine (and the rhizomes) are potentially mutagenic, so that they should be used with caution (e.g. not during pregnancy).

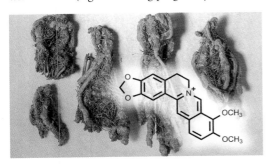

Coptis chinensis Franch. (Ranunculaceae); *huang lian* (Chinese); *huánglián, coptide chinois* (French); *Chinesischer Goldfaden* (German); *cottide* (Italian)

Coriandrum sativum
coriander

CLASSIFICATION TM: Asia, Europe. Pharm., Comm.E+, PhEur8.

USES & PROPERTIES Ripe fruits (*Coriandri fructus*) or their essential oil (*Coriandri aetheroleum*) are traditionally used against loss of appetite and as a stomachic to treat minor digestive disturbances, including indigestion, bloating and griping. They can also be used topically on wounds and burns and as counter-irritant on painful joints.

ORIGIN Eastern Mediterranean region and western Asia. Cultivated as an important culinary herb (cilantro, Chinese parsley) and spice (coriander, much used in curry powder).

BOTANY Annual herb (to 0.5 m); leaves dimorphic (basal ones undivided, upper much dissected; flowers asymmetrical; fruit a small dry schizocarp.

CHEMISTRY Essential oil: **linalool** (60%) as main compound. Also coumarins and triterpenoids.

PHARMACOLOGY The fruits (and essential oil) are spasmolytic, carminative and antimicrobial.

TOXICOLOGY Edible (but allergic skin reactions may occur in sensitive persons).

Coriandrum sativum L. (Apiaceae); *coriandre* (French); *Koriander* (German); *coriandolo* (Italian); *cilantro* (Spanish)

Crataegus monogyna
hawthorn

CLASSIFICATION TM: Europe, Asia. Comm.E+, ESCOP 6, WHO 2, 5, PhEur8, clinical studies+.

USES & PROPERTIES Flowers and leaves (*Crataegi folium cum flore*) are used in traditional medicine as heart tonic to treat cardiac insufficiency and heart rhythm disorders (NYHA I, II or III). The related *C. laevigata* (Europe) and *C. pinnatifida* (China) are also used. Special extracts are used in modern phytotherapy.

ORIGIN Europe, Asia (*C. laevigata* in Europe).

BOTANY Deciduous small tree; leaves deeply lobed; fruits fleshy, single-seeded (leaves shallowly lobed and fruits 2–3-seeded in *C. laevigata*).

CHEMISTRY Oligomeric procyanidins (1–3%), flavonoids (1–2%, e.g. **vitexin** rhamnoside), organic acids and triterpenes.

PHARMACOLOGY Cardiotonic activity is due to the procyanidins and flavonoids: their hydroxy groups interfere with enzymes and increase the strength of contraction and stroke volume of the heart. There are also anti-arrhythmic effects.

TOXICOLOGY Safe at recommended doses.

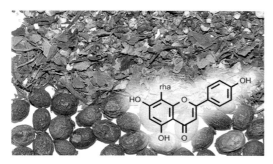

Crataegus monogyna Jacq. (Rosaceae); *aubépine* (French); *Eingriffeliger Weißdorn* (German); *bianco spino* (Italian); *espino albar* (Spanish)

Crocus sativus
saffron • saffron crocus

CLASSIFICATION Neurotoxin, mind-altering (II). TM: Europe, Asia. Pharm., Comm.E+, WHO 3.

USES & PROPERTIES The dried stigmas and style branches of the flowers (*Croci stigma*) are traditionally used to treat spasms and asthma. It is a component of "Swedish bitters". Saffron is the most expensive of spices and provides both colour and flavour to many Mediterranean and Middle Eastern traditional dishes (e.g. paella). About 150 000 flowers are required to produce 1 kg of dry spice.

ORIGIN A sterile cultigen developed in ancient Greece. Production is centred in Spain and Iran.

BOTANY Small deciduous bulbous plant; leaves slender, needle-shaped; flowers purple; stamens 3; stigmas bright red, on forked style branches.

CHEMISTRY Main compounds: **crocetin** (a diterpene) and picrocrocin (a glycoside). The essential oil contains safranal as the main active constituent.

PHARMACOLOGY Saffron has sedative and antispasmodic activities. Crocetin is lipid-lowering.

TOXICOLOGY Saffron is toxic at high doses. The lethal oral dose in humans is 5–20 g.

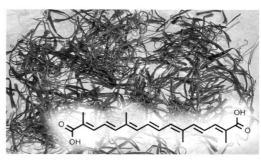

Crocus sativus L. (Iridaceae); *safran* (French); *Safran* (German); *zafferano* (Italian); *azafrán* (Spanish)

Croton tiglium
purging croton

CLASSIFICATION Cell toxin, skin irritant, extremely hazardous (Ia). TM: Europe (obsolete).

USES & PROPERTIES The seeds or seed oil were once popular and effective purgative medicines (no longer used due to toxicity and carcinogenic effects). Laxative medicines are less important now because of improved food hygiene and healthier diets. Habitual purgation may be harmful.

ORIGIN Asia (China, India, Malaysia).

BOTANY Shrub or small tree; leaves with two small glands; flowers small; fruit a dehiscent capsule; seeds marbled (resembling castor beans).

CHEMISTRY Phorbol esters of the tigliane type, such as TPA or 12-O-tetradecanoylphorbol-13-acetate; seeds contain fatty acids (**crotonic acid**), crotonide (purine alkaloid, 3.8%) and a toxic lectin (crotin).

PHARMACOLOGY Phorbol esters are co-carcinogens and extremely toxic. The purgative action is due to crotonic acid. Crotin is similar to ricin.

TOXICOLOGY Lethal dose in humans: four seeds or 0.5–1 ml (20 drops) of seed oil.

Croton tiglium L. (Euphorbiaceae); *croton revulsif* (French); *Krotonölbaum* (German); *crotone* (Italian)

Cucurbita pepo
pumpkin

CLASSIFICATION TM: Europe. Pharm., Comm. E+, ESCOP, HMPC.

USES & PROPERTIES Ripe, dried pumpkin seeds (*Cucurbitae peponis semen*) or seed oil are traditionally used in central Europe as vermifuge (to expel tapeworms and roundworms) and to treat the symptoms of benign prostate hyperplasia. The daily dose is 10 g but treatment should be maintained for several weeks or months to be effective.

ORIGIN Central and South America. Now a popular vegetable in most parts of the world.

BOTANY Trailing annual; leaves large, hairy; flowers yellow, short-lived; fruit fleshy, many-seeded.

CHEMISTRY The seed oil contains fatty acids (mainly linoleic acid), plant sterols and sterol glycosides (Δ^7-sterols) and tocopherols. **β-sitosterol** is a main compound. Also present is cucurbitine (a cyclic non-protein amino acid).

PHARMACOLOGY Δ^7-sterols inhibit the binding of dihydrotestosterone and/or 5α-amilase and aromatase. Cucurbitine is thought to be anthelmintic.

TOXICOLOGY Phytosterols are not poisonous.

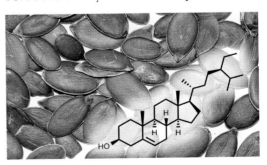

Cucurbita pepo L. (Cucurbitaceae); *pépon, citrouille* (French); *Gartenkürbis* (German); *zucca* (Italian); *calabaza* (Spanish)

Curcuma longa
turmeric

CLASSIFICATION TM: Asia, Europe. Comm.E+, WHO 1, ESCOP Suppl., PhEur8, HMPC.
USES & PROPERTIES The rhizomes (*Curcumae longae rhizoma*) are used in Ayurvedic medicine for numerous ailments, including indigestion and inflammatory conditions. It is used as a cholagogue (to stimulate bile flow) and as carminative (to reduce bloating). Turmeric (fresh or dried) is an important spice and natural food dye.
ORIGIN Asia (a sterile cultigen, thought to have originated in India). The spice is grown commercially in tropical areas around the world.
BOTANY Leafy perennial herb; rhizomes bright yellow; leaves large; flowers yellow and white.
CHEMISTRY The yellow pigments are curcuminoids (**curcumin** is the main compound). The essential oil contains bisabolane.
PHARMACOLOGY Curcumin and related compounds are anti-inflammatory, antioxidative, antimicrobial and cytotoxic to tumour cells. They are thought to have choleretic activity.
TOXICOLOGY Turmeric is non-toxic (edible).

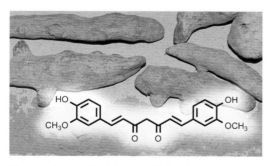

Curcuma longa L. [= *C. domestica* Valeton] (Zingiberaceae); *curcuma* (French); *Kurkuma, Gelbwurzel* (German); *curcuma* (Italian); *turmérico* (Spanish)

Cyamopsis tetragonolobus
cluster bean • guar

CLASSIFICATION TM: Asia, Europe. Pharm. DS: prebiotic (dietary fibre).
USES & PROPERTIES Seeds or seed oil are used in Ayurvedic medicine (and in Europe) as traditional dietary supplements and functional food, in supportive treatment of digestive ailments, constipation and diabetes. The seed gum (guar gum) is an industrial food additive, stabiliser and emulsifier. The gum or partially hydrolysed guar gum (PHGG) has health benefits as a dietary fibre.
ORIGIN A cultigen (unknown in the wild), probably originating from West African *C. senegalensis*. Cultivated in India and Pakistan for centuries.
BOTANY Erect annual herb; leaves trifoliate; flowers white or pink; fruit an oblong, 10-seeded pod.
CHEMISTRY Polysaccharides: mainly guaran, a galactomannan in the endosperm (MW 25 000, with **mannose** and galactose subunits).
PHARMACOLOGY The gum is considered to be a prebiotic (non-digestible fibre that acts as a substrate for beneficial microorganisms in the colon).
TOXICOLOGY The seeds and gum are edible.

Cyamopsis tetragonolobus (L.) Taubert (Fabaceae); *guar* (French); *Guarbohne, Büschelbohne* (German); *guar, guwar, guar-phali* (Hindi); *guar* (Italian); *guar* (Spanish)

Ecballium elaterium
squirting cucumber

CLASSIFICATION Cell toxin, highly hazardous (Ib); TM: Europe.
USES & PROPERTIES The fruits have been used in former times as a drastic purgative and anti-inflammatory medicine.
ORIGIN Europe (Mediterranean), Australia and western Asia (introduced to Central America).
BOTANY Trailing perennial herb; leaves hairy; flowers yellow; fruits ellipsoid and bristly. The plant is called squirting cucumber because the content of the fruit comes under pressure as the fruit ripens and is then forcefully ejected.
CHEMISTRY Cucurbitacins (bitter triterpenoids): **cucurbitacin I** and E are examples. The yield is up to 2.2% of fresh weight.
PHARMACOLOGY Cucurbitacins are extremely bitter and very poisonous. Some of them inhibit cell division and are therefore cytotoxic (but too toxic to use for their antitumour activity).
TOXICOLOGY Extremely poisonous: 0.6 ml of fruit juice can be lethal to humans. Cucurbitacin B: LD_{50} = 5 mg/kg (mouse, p.o.).

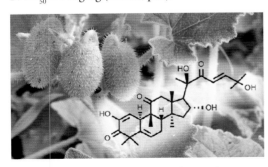

Ecballium elaterium (L.) A. Rich (Cucurbitaceae); *concombre sauvage* (French); *Spritzgurke* (German); *cocomero asinino* (Italian)

Echinacea pallida
pale purple coneflower

CLASSIFICATION TM: North America, Europe. Comm.E+, ESCOP 6, WHO 1, PhEur8, HMPC, clinical studies+.
USES & PROPERTIES The fresh or dried roots (*Echinaceae pallidae radix*) are used as general tonic and immune stimulant, for supportive treatment of colds and influenza.
ORIGIN North America (central parts). Roots of *E. angustifolia* (top right) have similar properties.
BOTANY Perennial herb (to 0.8 m); leaves narrow, basal; flower heads pale purple, on long stalks.
CHEMISTRY The roots contain polysaccharides, alkamides (e.g. echinaceine), polyacetylenes (e.g. ponticaepoxide) and antioxidant caffeic acid derivatives (e.g. **cynarin**, unique to *E. pallida*).
PHARMACOLOGY The product is considered to have immune stimulant activity (ascribed to the polysaccharides, based on *in vitro* studies only). Various compounds possibly act in synergy. Clinical studies on flu patients showed a reduction in recovery time.
TOXICOLOGY Safe to use at recommended doses.

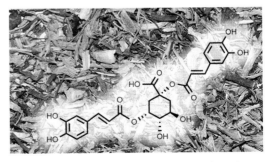

Echinacea pallida (Nutt.) Nutt. (Asteraceae); *échinacée* (French); *Blasser Sonnenhut* (German); *rudbeckia, pigna rossa* (Italian)

Echinacea purpurea
echinacea • purple coneflower

CLASSIFICATION TM: North America, Europe. Comm.E+, ESCOP 6, WHO 1, PhEur8, HMPC, clinical studies+.
USES & PROPERTIES The whole herb (*Echinaceae purpureae herba*), root (*Echinaceae purpureae radix*) or juice from fresh herb (6–9 ml per day) are used in supportive treatment of colds and infections of the respiratory and urinary tract. Externally, it is used to treat wounds and ulcers.
ORIGIN North America (widely cultivated).
BOTANY Perennial herb (to 1 m); leaves broad, bristly; flower heads purple, on short stalks.
CHEMISTRY The products contain characteristic polysaccharides, caffeic acid derivatives (with **cichoric acid** as main compound), polyacetylenes and akylamides (isobutylamides, e.g. echinaceine).
PHARMACOLOGY The immune stimulant activity is ascribed to polysaccharides. Chicoric acid and echinaceine are probably antimicrobial and anti-inflammatory. Clinical studies showed a reduction of infections and cold symptoms.
TOXICOLOGY No serious side effects are known.

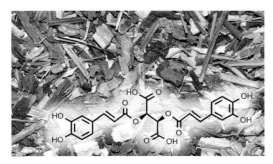

Echinacea purpurea (L.) Moench (Asteraceae); *échinacée, rudbeckie poupre* (French); *Purpur-Sonnenhut* (German); *rudbeckia rossa, echinacea* (Italian)

Elettaria cardamomum
cardamom

CLASSIFICATION TM: Asia (India), Europe. Pharm., Comm.E+, WHO 4.
USES & PROPERTIES Fruits and seeds (*Cardamomi fructus*) or seed oil (*Cardamomi aetheroleum*) are used, rarely the rhizomes. Seeds are an important spice but also an aphrodisiac in Ayurvedic medicine: used to treat bad breath, respiratory ailments (cough, asthma, bronchitis), digestive problems, as reputed cholagogue (for nausea, griping, stomach pain, flatulence) and urinary complaints.
ORIGIN Asia (India and Sri Lanka). Cultivated in tropical countries (India, Indonesia, Malaysia).
BOTANY Leafy perennial herb; fruit a many-seeded, 3-valved capsule; seeds brown, warty.
CHEMISTRY The seeds contain essential oil (4%) with **1,8-cineole** as main compound.
PHARMACOLOGY The essential oil has antimicrobial and spasmolytic activities. Extracts were shown to stimulate the excretion of bile and gastric juices in animals.
TOXICOLOGY Non-toxic (edible).

Elettaria cardamomum (L.) Maton (Zingiberaceae); *cardamomier* (French); *Grüner Kardamom, Kardamompflanze* (German); *cardamomo* (Italian); *ela* (Sanskrit)

Eleutherococcus senticosus
Siberian ginseng • eleuthero

CLASSIFICATION TM: Asia; Comm.E+, ESCOP, WHO 2, PhEur8, HMPC.

USES & PROPERTIES Dried rhizomes and roots (*Eleutherococci radix*) are used as adaptogenic tonic to reduce stress and fatigue, to improve convalescence and to reverse the symptoms of age-related decreases in physical and mental capacity.

ORIGIN Northeastern Asia (eastern Siberia).

BOTANY Woody shrub; stems spiny; leaves compound; flowers small, arranged in umbels.

CHEMISTRY Complex: coumarins (isofraxidin), lignans and their glycosides (sesamin), phenylpropanoids, triterpene saponins (eleutherosides I–M). An example of a lignan is **eleutheroside D**.

PHARMACOLOGY Animal and human studies indicated improved endurance and stress resistance, similar to real ginseng (see *Panax ginseng*).

TOXICOLOGY Treatment for three months maximum; unsafe for those with high blood pressure.

NOTES An adaptogen has both reducing and increasing biological activities and restores the equilibrium that was disrupted by disease.

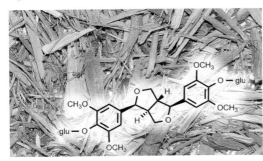

Eleutherococcus senticosus (Rupr. & Maxim.) Maxim. (Araliaceae); *éleuthérocoque, ginseng de Sibérie* (French); *Stachelpanax, Sibirischer Ginseng* (German)

Elymus repens
couchgrass • twitch

CLASSIFICATION TM: Europe. Comm.E+, PhEur8.

USES & PROPERTIES The rhizomes (*Agropyri repentis rhizoma* or *Graminis rhizoma*) are a traditional treatment for inflammation of the respiratory and urinary tracts (to prevent kidney gravel and to alleviate the symptoms of cystitis, urethritis and prostatitis (also gout and rheumatism). Commercial preparations or decoctions (4–9 g of dry rhizome per day) can be taken.

ORIGIN Europe, Asia, North and South America. A cosmopolitan grass and weed.

BOTANY Perennial grass (to 1 m); spikelets small, in two rows on slender unbranched spikes.

CHEMISTRY Polyfructosans: **triticin** as a main compound; essential oil: carvacrol, thymol, carvone); also mucilage, saponins and silica.

PHARMACOLOGY Polysaccharides are often used as diuretics (activity not yet explained). Mucilages have a soothing effect; the essential oil compounds are antimicrobial and perhaps also diuretic.

TOXICOLOGY The herb is safe to use.

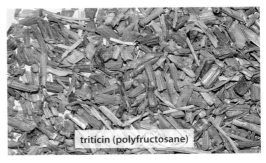

Elymus repens (L.) Gould [= *Agropyron repens* (L.) P.Beauv.] (Poaceae); *chiendent officinal* (French); *Gemeine Quecke* (German); *gramigna canina* (Italian)

Ephedra sinica
ephedra • desert tea

CLASSIFICATION Mind-altering, moderately hazardous (II). TM: Asia (China). Pharm., Comm. E+, WHO 1, PhEur8. MM: ephedrine.
USES & PROPERTIES The dried stems (*Ephedrae herba*; *ma huang*) are mostly used, rarely the roots (*Ephedrae radix*; *ma huang gen*). In Chinese traditional medicine, *ma huang* is mainly used to treat bronchitis, asthma and nasal congestion (rhinitis and sinusitis). Nowadays it is a popular ingredient of weight loss products and formulations aimed at improving the performance of athletes.
ORIGIN Asia and East Asia (China).
BOTANY Leafless shrub (to 1 m); stems silvery, ribbed; cones fleshy, bright red. Several other species have been used as a source of raw material.
CHEMISTRY Phenylethylalkaloids: **ephedrine** (produced by all *Ephedra* species).
PHARMACOLOGY Ephedrine is a central stimulant; it is also analeptic and bronchodilatory.
TOXICOLOGY Long-term use may lead to dependency. Ephedrine: lethal dose = 4.3 g/kg (mouse, p.o); banned as a dope in sports events.

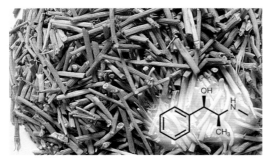

Ephedra sinica Stapf (Ephedraceae); *ma huang* (Chinese); *ephedra, raisin de mer* (French); *Ephedra, Meerträubel* (German); *efedra, uva marina* (Italian)

Equisetum arvense
horsetail • field horsetail

CLASSIFICATION TM: Europe, Asia. Pharm.
USES & PROPERTIES Dried stems (*Equiseti herba*) are traditionally used as haemostyptic (to reduce menstrual bleeding) and topically to treat slow healing wounds. Infusions of 2–4 g (6 g per day) are used as diuretic to treat urinary tract infections, kidney gravel and post-traumatic oedema.
ORIGIN Europe, Asia and North America. Raw material (sterile summer branches) are wild-harvested in eastern European countries.
BOTANY Perennial herb with seemingly leafless stems; unbranched fertile stems in early spring; sterile stems with many whorled nodes in summer.
CHEMISTRY The herb contains **silicic acid** (5–8%) and equisetolic acid (a dicarboxylic acid). Also potassium and aluminium salts, and flavonoids (glycosides of quercetin and kaempferol).
PHARMACOLOGY The diuretic and wound-healing activity is ascribed to the silica, potassium salts and the flavonoids.
TOXICOLOGY Safe, except when contaminated with potentially toxic marsh horsetail (*E. palustre*).

Equisetum arvense L. (Equisetaceae); *prêle de champs* (French); *Ackerschachtelhalm* (German); *coda di cavallo, equiseto dei campi* (Italian)

Erythroxylum coca
coca plant

CLASSIFICATION Neurotoxin, mind-altering (Ib). TM: South America; MM: alkaloids (derivatives).

USES & PROPERTIES The leaves (*Cocae folium*) are traditionally chewed in the Andean region to counteract fatigue and stress. Source of the alkaloid cocaine (the first commercial anaesthetic, later replaced by synthetic derivatives such as lidocaine). The modern illegal misuse of cocaine is for its stimulant and euphoric effects (thin lines are snorted from a flat surface through a straw).

ORIGIN South America (cultivated since ancient times. *E. coca* occurs in Bolivia and Peru; *E. novogranatense* in Colombia, Peru and Venezuela.

BOTANY Woody shrub (to 1 m); leaves simple, glabrous; flower white; fruit a few-seeded drupe.

CHEMISTRY Tropane alkaloids: mainly **cocaine**.

PHARMACOLOGY Cocaine is an illegal intoxicant with local anaesthetic and euphoric effects.

TOXICOLOGY Overdosing on cocaine is often fatal: 1–2 g (p.o.) or 0.2–0.3 g (s.c.); a mere 30 mg can be lethal in sensitive persons. Highly addictive, with strong psychic but no physical dependence.

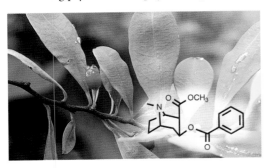

Erythroxylum coca Lam. (Erythroxylaceae); *cocalier* (French); *Kokastrauch* (German); *coca* (Italian); *cocal* (Spanish)

Eucalyptus globulus
eucalyptus • bluegum

CLASSIFICATION TM: Australia, Europe. Pharm., Comm.E+, ESCOP 6, WHO 2, PhEur8, HMPC.

USES & PROPERTIES Mature (upper) leaves (*Eucalypti folium*) or their essential oil (*Eucalypti aetheroleum*) are mainly used to treat ailments of the respiratory tract (including cough, colds, nasal congestions and bronchial infections). The oil is used topically to treat skin ailments, minor wounds and rheumatism. Small amounts (0.3–0.6 ml per day) can be safely ingested or a tea is made from 1.5–2 g of leaves in 150 ml water.

ORIGIN Australia; commercially cultivated in Spain, Morocco and many other countries.

BOTANY Evergreen tree (to 60 m); mature leaves sickle-shaped, grey; flowers large, solitary, white.

CHEMISTRY Essential oil: **1,8-cineole** is the main compound (also known as eucalyptol). In addition, flavonoids and sesquiterpenes.

PHARMACOLOGY The essential oil is antiseptic and expectorant. The antimicrobial activities of 1,8-cineole and camphor are synergistic.

TOXICOLOGY Safe to use at recommended doses.

Eucalyptus globulus Labill. (Myrtaceae); *eucalyptus* (French); *Eukalyptus, Blaugummibaum* (German); *eucalipto* (Italian); *eucalypto* (Spanish)

Eucommia ulmoides
eucommia • Chinese rubber tree

CLASSIFICATION TM: East Asia (China).
USES & PROPERTIES Tea made from the inner bark (*Eucommiae cortex*) is used in Chinese medicine as a kidney and liver tonic and especially as a remedy for high blood pressure. The daily dose is 6–9 g. It is used (often with other Chinese herbs) for pain, sedation and inflammation.
ORIGIN East Asia (endemic to China).
BOTANY Deciduous tree; leaves simple, with latex threads when broken; fruits winged.
CHEMISTRY Iridoid glucosides (especially **geniposidic acid**); lignans (mainly pinoresinol diglucoside), flavonoids and organic acids. The latex/rubber (gutta-percha) content is 6–10%.
PHARMACOLOGY Geniposidic acid reduces blood pressure (and can be used as health food additive for prehypertension). The antioxidant lignan is also claimed to be hypertensive (pinoresinol may be hypoglycaemic). Bark has modest anti-inflammatory activity.
TOXICOLOGY The bark tea is used daily by elderly people in China, with no apparent ill effects.

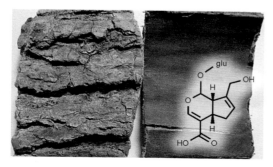

Eucommia ulmoides Oliv. (Eucommiaceae); *du zhong* (Chinese); *Gummiulme* (German)

Euonymus europaeus
spindle tree • European spindle

CLASSIFICATION Heart toxin, highly hazardous (Ib). TM: Europe, North America.
USES & PROPERTIES Root and stem bark of *Euonymus* species was formerly used to treat ailments of the liver and gall bladder, as well as skin disorders. The timber has many traditional uses (e.g. to make charcoal for gunpowder; violin bows) and the seeds contain a yellow dye (carote-noids) used to colour butter.
ORIGIN Europe and western Asia.
BOTANY Deciduous shrub or small tree (to 6 m); twigs four-angled; flowers small, greenish-white; fruit a dehiscent capsule; seeds orange, arillate.
CHEMISTRY Cardiac glycosides (cardenolides) such as evonoside, evobioside and **evomonoside**; toxic alkaloids (e.g. evonine); tannins and lectins.
PHARMACOLOGY The cardenolides inhibit Na^+, K^+-ATPase, resulting in cardiac arrhythmia, tachycardia, coma and death.
TOXICOLOGY Evonoside: LD_{50} = 0.84 mg/kg (cat, i.v.). Two fruits can cause severe poisoning in children; 36 are said to be lethal in adults.

Euonymus europaeus L. (Celastraceae); *fusain d'Europe* (French); *Gewöhnliches Pfaffenhütchen* (German); *fusaria commune* (Italian)

Euphorbia peplus
petty spurge • milkweed

CLASSIFICATION Cell toxin (II). TM: Europe, Africa, Asia. MM: diterpene.
USES & PROPERTIES All plant parts contain latex, used to treat skin ailments and cancerous ulcers. A pharmaceutical-grade ingenol mebutate skin gel to treat actinic keratosis has been approved by the US Food and Drug Administration and the European Medicines Agency.
ORIGIN North Africa, Europe and western Asia. A weedy plant of gardens and disturbed places. Naturalised and invasive in many regions, including North America, Australia and New Zealand.
BOTANY Annual herb (to 0.3 m) with white latex; leaves hairless, simple; flowers yellowish green.
CHEMISTRY Diterpene esters (in latex): mainly **ingenol mebutate** (ingenol-3-angelate).
PHARMACOLOGY Ingenol mebutate gel applied for 2–3 days is effective against solar keratoses (premalignant skin lesions caused by the sun).
TOXICOLOGY *Euphorbia* species are potentially lethal (they contain toxic phorbol esters). Topical use may cause irritation which can be severe.

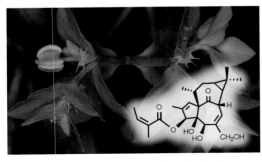

Euphorbia peplus L. (Euphorbiaceae); *euphorbe des jardiniers* (French); *Gartenwolfsmilch* (German)

Euterpe oleracea
acai berry

CLASSIFICATION DS: South America.
USES & PROPERTIES Ripe fruits (and juice) are a traditional Amazon food item, marketed since 2004 as açai, a supplement and health drink (with claims of efficacy in weight loss).
ORIGIN South America (Panama to Brazil). Wild-harvested but also cultivated commercially.
BOTANY Multi-stemmed palm (to 20 m); leaves pinnately dissected (3 m long); fruit a small drupe, purplish black or black (green in one cultivar).
CHEMISTRY The fruit pulp contains moderate quantities of anthocyanins: mainly **cyanidin-3-*O*-glucoside** (with several others). Also present are numerous flavonoids and low levels of resveratrol.
PHARMACOLOGY Anthocyanins: antioxidant and free radical scavenging activity. Convincing scientific evidence for claimed health benefits (and efficacy in weight loss) of açai is not yet available. The antioxidant activity of the juice is lower than that of pomegranate, black grapes and blueberries.
TOXICOLOGY The fruits are used as food and can be safely consumed.

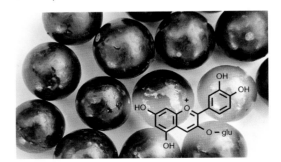

Euterpe oleracea Mart (Arecaceae); *Kohlpalme, Açaí-Beere* (German); *açaizeiro* (Portuguese)

Fabiana imbricata
Chilean false heath • pichi

CLASSIFICATION Neurotoxin, mind-altering (II). TM: South America, Europe.
USES & PROPERTIES The twigs and leaves were traditionally used in Chile and in Europe as diuretic and to treat ailments of the kidneys and urinary tract, including cystitis and gonorrhoea. South American Indians inhaled smoke from the dried twigs as an intoxicant.
ORIGIN South America (cultivated as a garden shrub in many parts of the world).
BOTANY Evergreen shrub (to 2 m); leaves ericoid; flowers tubular, pale purple or white.
CHEMISTRY Quinoline alkaloids: **fabianine**. Also present are essential oil, anthraquinones (physcione), sesquiterpenes (3,11-amorphadien and others), coumarins and flavonoids.
PHARMACOLOGY Fabianine can modulate neuroreceptors, resulting in euphoria. Extracts have antibacterial, diuretic and bitter tonic effects.
TOXICOLOGY Fabianine has a very low toxicity (up to 5 g extract per kg body weight had limited effects on rats).

Fabiana imbricata Ruiz & Pav. (Solanaceae); *fabiane imbriquée* (French); *Fabiane* (German); *pichi-pichi* (Italian); *pitchi, pichi romero* (Spanish)

Fagopyrum esculentum
buckwheat

CLASSIFICATION TM: Europe. Pharm., PhEur8. MM: rutin.
USES & PROPERTIES Dried aerial parts (*Fagopyri herba*) are traditionally used to treat the symptoms of capillary and venous insufficiency (bleeding, bruising, haemorrhoids, retinal haemorrhage, varicose veins and poor circulation).
ORIGIN Central and northern Asia. Buckwheat is widely cultivated as a food plant and is also a natural source of rutin (up to 8% in special cultivars).
BOTANY Annual herb (to 0.7 m); leaves heart-shaped; flowers white; fruits small, angular nutlets.
CHEMISTRY Flavonoids (so-called bioflavonoids): mainly **rutin** (quercetin-3-rutinoside). Flowers and seed husks contain fagopyrine (a diathrone).
PHARMACOLOGY Rutin and other antioxidant flavonoids have venotonic and vascular protective activity. They improve the elasticity of veins and enhance circulation.
TOXICOLOGY Aerial parts are non-toxic in small doses. Buckwheat has become popular as a functional food and dietary supplement.

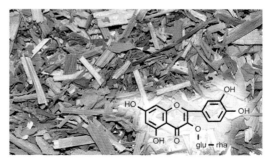

Fagopyrum esculentum Moench (Polygonaceae); *sarrasin* (French); *Echter Buchweizen* (German); *grano saraceno* (Italian)

Gaultheria procumbens
wintergreen • checkerberry

CLASSIFICATION TM: North America, Europe.
USES & PROPERTIES Leaves (*Gaultheriae folium*), as infusions or decoctions, and small amounts of essential oil (*Gaultheriae aetheroleum*) are taken orally as tonics and stomachics, but also for rheumatism. Wintergreen oil and ointments are counter-irritants for treating painful joints and muscles. It is also an ingredient of oral hygiene products.
ORIGIN North America (Canada and eastern parts of the USA). The product is mainly wild-harvested.
BOTANY Evergreen, mat-forming shrub; leaves leathery; flowers white or pinkish; fruit edible.
CHEMISTRY Essential oil: almost pure **methyl salicylate** (it occurs in the plant as a glycoside). Leaves contain arbutin and tannins.
PHARMACOLOGY Methyl salicylate, released by hydrolysis of the glycoside in the intestine and liver, has anti-inflammatory and analgesic activities. Arbutin and tannins are also anti-inflammatory.
TOXICOLOGY Use only in small amounts. The lethal dose in adult humans is 30 g of the oil.

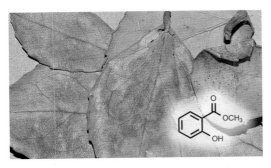

Gaultheria procumbens L. (Ericaceae); *gaulthérie du Canada, thé des bois* (French); *Niederliegende Scheinbeere, Wintergrün* (German); *uva di monte* (Italian)

Gelsemium sempervirens
yellow jasmine

CLASSIFICATION Neurotoxin (Ia). TM: North America. Comm.E.
USES & PROPERTIES Rhizomes and roots (*Gelsemii rhizoma*) are used to extract alkaloids for the treatment of facial and dental neuralgia (intense pain cause by damaged nerves). Included in cough syrups (for asthma and whooping cough) and topically against the pain caused by a pinched nerve. Fresh rhizomes are used in homoeopathy for treating migraine, anxiety and dysmenorrhoea.
ORIGIN North America (southeastern USA). An attractive and popular ornamental plant.
BOTANY Evergreen climber (vine); leaves opposite, glossy green; flowers bright yellow, tubular.
CHEMISTRY Monoterpene indole alkaloids (0.5%): **gelsemine** (the main compound); also coumarins, iridoids and pregnane-type steroids.
PHARMACOLOGY Gelsemine modulates the glycine receptor (similar to strychnine): it is antispasmodic, analgesic and sedative.
TOXICOLOGY Gelsemine is very poisonous. LD_{50} (mouse) = 4 mg/kg (i.v.) or 1.24 g/kg (p.o.).

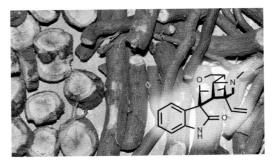

Gelsemium sempervirens (L.) J. St-Hil. (Gelsemiaceae); *jasmin sauvage* (French); *Falscher Jasmin, Giftjasmin* (German); *gelsemino* (Italian)

Gentiana lutea
yellow gentian

CLASSIFICATION TM: Europe. Comm.E+, ESCOP 4, WHO 3, PhEur8, HMPC, clinical studies+.

USES & PROPERTIES The rhizome and root (*Gentianae radix*) are traditionally used as digestive bitter tonic, cholagogue and stomachic to stimulate appetite and to treat dyspepsia with anorexia. It is used in commercial herbal tinctures, liqueurs and roborants, and in homeopathic preparations.

ORIGIN Europe (usually at high elevations).

BOTANY Deciduous perennial herb (to 1.5 m); leaves with prominent veins; flowers yellow, in clusters on a long stalk.

CHEMISTRY Bitter secoiridoids: mainly gentiopicroside (=**gentiopicrin**) (2–3%). The bitter taste is due to a minor compound, amarogentin (bitterness value of 50 000 000). Xanthones (gentisin, gentioside) give the yellow colour to the roots.

PHARMACOLOGY The bitter compounds stimulate (via the *nervus vagus*) the flow of saliva, gastric juices and bile. Gentian root also has antimicrobial, anti-stress and immune-modulatory activities.

TOXICOLOGY Non-toxic at low doses.

Gentiana lutea L. (Gentianaceae); *gentiane jaune* (French); *Gelber Enzian* (German); *genziana maggiore* (Italian)

Ginkgo biloba
ginkgo • maidenhair tree

CLASSIFICATION TM: Asia (China). Comm.E+ (extracts only), ESCOP, WHO 1, PhEur8, clinical studies+.

USES & PROPERTIES The leaves (and seeds) have many traditional uses but special leaf extracts (120–240 mg/day) are nowadays used for treating cerebrovascular insufficiency and symptoms of old age (e.g. sleep disturbances, memory loss, dementia and peripheral arterial occlusive disease).

ORIGIN Eastern Asia (China). A popular street tree, commercially cultivated for leaves and nuts.

BOTANY Large tree (to 35 m); leaves bilobed, with striate venation; fruit fleshy; seed (nut) edible.

CHEMISTRY Special extracts are almost devoid of ginkgolic acids but rich in flavonoids (flavonol glycosides and biflavonoids), unique diterpene lactones (**ginkgolide B** and others), a sesquiterpenoid (bilobalide) and oligomeric proanthocyanidins.

PHARMACOLOGY Clinical studies support the efficacy of ginkgo extracts in treating circulatory disorders, tinnitus and dementia.

TOXICOLOGY Safe at prescribed doses.

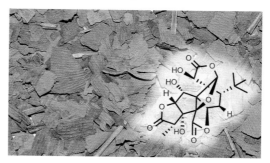

Ginkgo biloba L. (Ginkgoaceae); *ginkgo* (French); *Ginkgo* (German); *ginkgo biloba* (Italian); *arbol de los escudos* (Spanish)

Gloriosa superba
flame lily • climbing lily

CLASSIFICATION Cell toxin, extremely hazardous (Ia). MM: source of colchicine.

USES & PROPERTIES The rhizomes have been used in traditional medicine (e.g. as abortifacient) but are nowadays mainly used for the extraction of colchicine. The pure alkaloid is used in cytological research and to treat inflammatory conditions (e.g. gout, see *Colchicum autumnalis*).

ORIGIN Africa (southern and eastern parts) to South Asia (India). An attractive garden plant.

BOTANY Geophyte; rhizomes fleshy; leaves with climbing tendrils; flowers spectacular, in shades of yellow and red; fruit a capsule; seeds orange.

CHEMISTRY Phenethylisoquinoline alkaloids: **colchicine** is the main compound (2.4% in leaves).

PHARMACOLOGY Colchicine is a well-known spindle poison that disrupts cell division. It has anti-inflammatory and analgesic activities.

TOXICOLOGY Colchicine is extremely poisonous: oral intake of more than 40 mg of the alkaloid results in death within three days, due to fatal respiratory and cardiovascular arrest.

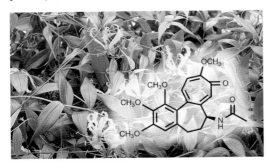

Gloriosa superba L. (Colchicaceae); *superbe de Malabar, lis de Malabar* (French); *Ruhmeskrone* (German)

Glycine max
soybean

CLASSIFICATION TM: Europe, Asia (China). Pharm., Comm.E+, PhEur8. DS: lecithin.

USES & PROPERTIES Soy seeds (*Sojae semen*), lecithin and oil (*Lecithinum ex soja*; *Sojae oleum*) are popular dietary supplements used to reduce cholesterol and the symptoms of menopause. Soy lecithin is used for appetite loss, chronic hepatitis, inflammatory bowel disease and acne. It is linked to the low incidence of cancer in East Asian people.

ORIGIN Asia. An old Chinese cultigen.

BOTANY Erect annual herb; leaves trifoliate; flowers minute, pink; fruit a few-seeded pod.

CHEMISTRY Soybean lecithin is a mixture of phospholipids (mainly **phosphatidylcholine**). Soy also has isoflavonoids (3 mg/g, genistein and daidzein), saponins (soyasaponins), omega-3 fatty acids (α-linolenic acid) and lunasin (a peptide).

PHARMACOLOGY Lecithin is a source of choline, an essential nutrient. It lowers blood lipids. Isoflavonoids are phytoestrogenic. Soyasaponins may be antithrombotic and liver-protectant.

TOXICOLOGY Allergy to soy is quite common.

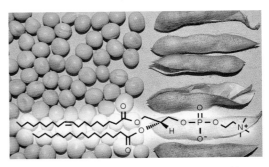

Glycine max (L.) Merr. (Fabaceae); *fève de soja* (French); *Sojabohne* (German); *soia* (Italian)

Glycyrrhiza glabra
liquorice • licorice

CLASSIFICATION TM: Europe, Asia (India). Pharm., Comm.E+, ESCOP, WHO 1, PhEur8, HMPC.

USES & PROPERTIES Dried rhizomes (*Liquiritiae radix*) are traditionally used (as an infusion of 1–1.5 g in 150 ml water) for coughs, catarrh, gastritis, flatulence and chronic gastric and duodenal ulcers. Externally it is applied for relief of sunburn, insect bites, pruritus and piles. It is included in products as active ingredient and as sweetener.

ORIGIN Mediterranean to Central Asia. *Gan cao* or Chinese liquorice (*G. uralensis*) is also used.

BOTANY Perennial herb (to 1 m); rhizomes and stems woody; leaves pinnate; flowers white/purple.

CHEMISTRY Flavonoids (mainly liquiritin); triterpene saponins: glycyrrhizic acid (2–15%), and its aglycone, **glycyrrhetinic acid**.

PHARMACOLOGY The saponins are expectorant, secretolytic and anti-inflammatory. Numerous other activities are associated with liquorice.

TOXICOLOGY Chronic use of large doses: mineralocorticoid effects and high blood pressure.

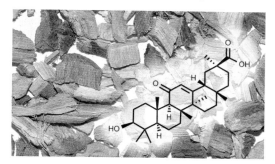

Glycyrrhiza glabra L. (Fabaceae); *réglisse officinale* (French); *Süßholz, Lakritze* (German); *liquirizia* (Italian); *regalicia* (Spanish)

Griffonia simplicifolia
griffonia

CLASSIFICATION TM: West Africa. AHP, clinical studies+.

USES & PROPERTIES Roots, stems and leaves are used in African traditional medicine as poultices for wounds and as anti-emetics, aphrodisiacs and to treat kidney ailments. The main interest is in the seeds: a rich source of 5-hydroxytryptophan. It is used for treating neurological and psychiatric disorders (mostly depression, but also anxiety, insomnia, headaches, migraine and eating disorders). Initial dose: 50 mg, 3 × per day with meals.

ORIGIN West tropical Africa (mainly Ghana).

BOTANY Woody climber (vine); fruit an inflated pod; seeds disc-shaped, black, 15–20 mm in diam.

CHEMISTRY Seeds contain several indole alkaloids (14–17% of dry weight), of which **5-hydroxytryptophan** (5-HTP) is the dominant compound.

PHARMACOLOGY 5-HTP increases the levels of serotonin in the brain and central nervous system.

TOXICOLOGY Low toxicity. 5-HTP: LD_{50} = 243 mg/kg (rat, p.o.). Not suitable for self-medication, due to the possibility of severe side effects.

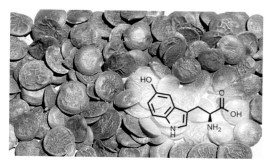

Griffonia simplicifolia (DC.) Baill. (Fabaceae); *griffonia* (French); *Griffonia* (German)

Heliotropium arborescens
heliotrope • cherry pie

CLASSIFICATION Liver toxin, neurotoxin, mutagen, moderately hazardous (II).

USES & PROPERTIES All parts of the plant contain pyrrolizidine alkaloids that may cause acute and more often chronic poisoning when ingested. Contamination of cereals with the seeds of *Heliotropium* species and related plants has been the cause of fatal human poisoning.

ORIGIN South America (Peru and Ecuador). Widely cultivated as garden shrub for the strong vanilla-like fragrance of the flowers.

BOTANY Perennial shrub (to 2 m); leaves hairy; flowers blue/purple, in scorpioid racemes.

CHEMISTRY The level of pyrrolizidine alkaloids (PAs) in aerial parts is about 0.9% of dry weight: **indicine** (main compound), with smaller amounts of acetylindicine, heliotropine and cynoglossine. Heliotropine is dominant in many species.

PHARMACOLOGY PAs with an unsaturated necine base are activated in the liver, causing damage.

TOXICOLOGY PAs are cumulative liver poisons; they are mutagenic, carcinogenic and teratogenic.

Heliotropium arborescens L. [= *H. peruvianum* L.] (Boraginaceae); *héliotrope, fleur des dames* (French); *Vanilleblume, Heliotrop* (German); *girasole del Peru* (Italian)

Helleborus viridis
green hellebore

CLASSIFICATION Heart toxin, extremely hazardous (Ia). TM: Europe (obsolete).

USES & PROPERTIES The roots and/or leaves were once used in traditional medicine to treat nausea, constipation, intestinal worms and nephritis. It was also employed as abortifacient. Powdered herb was formerly used in central Europe as sneezing powder (now banned).

ORIGIN Western and central Europe. Several species are cultivated ornamentals, including *H. niger*.

BOTANY Small perennial herb; rhizomes fleshy; leaves compound; flowers nodding, green (white in *H. niger*, photo below); seeds with arils.

CHEMISTRY Heart glycosides of the bufadienolide type (up to 1.5%): **hellebrin** is the main compound. Also present are alkaloids and saponins (and irritant ranunculin in *H. niger*).

PHARMACOLOGY In high doses, bufadienolides can cause fatal cardiac and respiratory arrest. The alkaloids resemble aconitine in their activity.

TOXICOLOGY The LD_{50} of helleborin is 8.4 mg/kg (mouse, i.p.) or a mere 0.1 mg/kg (cat, i.v.).

Helleborus viridis L. (Ranunculaceae); *hellébore vert* (French); *Grüne Nieswurz* (German); *elleboro verde* (Italian)

Herniaria glabra
smooth rupturewort • herniary

CLASSIFICATION TM: Europe. Comm.E+.
USES & PROPERTIES The dried, aboveground parts (rupturewort herb – *Herniariae herba*) is a traditional diuretic medicine, used to treat bladder and kidney ailments, including chronic cystitis, and urethritis. A decoction of 1.5 g is taken two or three times per day. It is a "blood purifier" that has also been used for respiratory tract infections, arthritis and rheumatism. The generic name suggests that the herb can heal hernias.
ORIGIN Europe and Asia (*H. glabra*) or the Mediterranean region and North Africa (*H. hirsuta*). Both species are acceptable as sources of the herb.
BOTANY Short-lived perennial herbs; stems mat-forming; leaves green, glabrous (or bluish green and hairy in *H. hirsuta*); flowers inconspicuous.
CHEMISTRY Mainly saponins (up to 9%): glycosides of **medicagenic acid** and gypsogenic acid. Also flavonols (1.2%), tannins and coumarins.
PHARMACOLOGY Weak spasmolytic activity is ascribed to the saponins and flavonoids.
TOXICOLOGY There is no evidence of side effects.

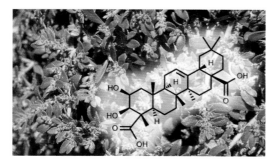

Herniaria glabra L. (Caryophyllaceae); *herniaire* (French); *Kahles Bruchkraut* (German); *erniaria* (Italian)

Hibiscus sabdariffa
hibiscus • red-sorrel • roselle

CLASSIFICATION TM: Africa, Europe. Pharm., Comm.E+, PhEur8, AHP. DS: functional food.
USES & PROPERTIES The dried calyces and epicalyces (hibiscus flowers – *Hibisci flos*) are used as a tasty, caffeine-free health tea and as a colourful additive to health tea mixtures. It has been used in traditional medicine as a general tonic to treat appetite loss, colds and catarrh. Externally it is said to help with skin ailments and allergic eczema.
ORIGIN Africa (Angola). Cultivated commercially in warm regions.
BOTANY Erect annual (to 4 m); leaves lobed; flowers white, epicalyx green, calyx fleshy, bright red (purple when dry), edible, with a sweet-sour taste.
CHEMISTRY Polysaccharides, pectins, organic acids (e.g. **hibiscus acid**), sugars, anthocyanins.
PHARMACOLOGY Some health benefits have been proposed. The anthocyanins are antioxidants, the organic acids are mild laxatives and the polysaccharides may have an immune-modulatory action and a soothing effect on inflamed mucosa.
TOXICOLOGY Non-toxic and edible.

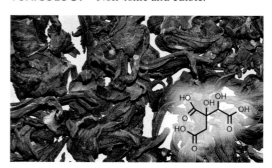

Hibiscus sabdariffa L. (Malvaceae); *karkadé* (French); *Hibiscus, Sabdariffa-Eibisch* (German); *carcade* (Italian)

Hippophae rhamnoides
buckthorn

CLASSIFICATION TM: Europe. WHO 5, clinical studies+. DS: functional food.

USES & PROPERTIES Ripe fruits (*Hippophae fructus*) have been used to treat asthma, gastric ulcers, and disorders of the lungs, liver and skin. Oil from the seeds and fruit pulp (*Hippophae oleum*) has become popular as a dietary supplement.

ORIGIN Europe and Asia.

BOTANY Deciduous, dioecious shrub or small tree (2–5 m); stems spiny; leaves narrow, with silver scales; fruit a bright orange drupe.

CHEMISTRY Fruits: vitamins C and E, carotenoids (lycopene and β-carotene), flavonoids and mannitol. Seed oil: linolenic and linoleic acids. Mesocarp oil: **palmitoleic acid** and palmitic acid (65%).

PHARMACOLOGY Clinical studies showed efficacy in reversing liver damage. Antioxidant, anti-inflammatory, hepatoprotective, neuroprotective, wound-healing, anti-ulcer, anticarcinogenic and antimutagenic activities have been recorded.

TOXICOLOGY No side effects are known.

Hippophae rhamnoides L. (Eleagnaceae) *argousier* (French); *Sanddorn* (German); *olivello spinoso* (Italian); *espino cerval de mar* (Spanish)

Hoodia gordonii
hoodia • ghaap

CLASSIFICATION TM: Africa. AHP. DS: functional food, clinical studies.

USES & PROPERTIES Pieces of stem were eaten by Khoi-San hunters and herders in South Africa and Namibia as a functional food and masticatory to suppress hunger and thirst. It became popular as a dietary supplement with claimed appetite-suppressant and anti-obesity activity.

ORIGIN Southern Africa. Found in arid regions.

BOTANY Leafless succulent (to 0.5 m); stems cactus-like, spiny; flowers flesh-coloured, foetid.

CHEMISTRY The main active compound is a pregnane glycoside called **P57**.

PHARMACOLOGY Appetite-suppressant activity was shown in animal experiments. Human clinical studies indicated an unfavourable risk to benefit ratio. The traditional method of administration (as masticatory) bypasses the stomach and the food aversion effect appears to be centred in the mouth (associated with an intense bitter taste).

TOXICOLOGY Side effects (at therapeutic doses) include nausea and hypertension.

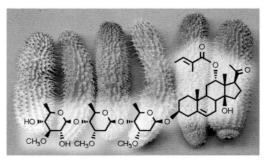

Hoodia pilifera (L.f.) Plowes [= *Trichocaulon piliferum* L.f.] [Apocynaceae (formerly Asclepiadaceae)]; *hoodia* (French); *Hoodia* (German); *hoodia* (Italian)

Hordeum vulgare
barley

CLASSIFICATION TM: Europe, Asia (China). DS: functional food, clinical studies+.

USES & PROPERTIES Barley is traditionally used to treat inflammation of the digestive tract. The grains and dried juice from seedlings (known as barley green) have become popular health food items for controlling diabetes, hyperlipidaemia, hypertension, coronary heart disease and gastrointestinal disorders. Promoting weight loss e.g. with pure hordenine) is not yet convincing.

ORIGIN Middle East. Barley was domesticated in Mesopotamia some 10 000 years ago.

BOTANY Annual grass (to 1 m); spikelets mostly in three rows; husk usually adhering to the grain.

CHEMISTRY Dietary fibre (β-glucans in the husk); essential nutrients, e.g. vitamin B5, B9 (folic acid), lysine; alkaloids (**hordenine**) in seedling roots.

PHARMACOLOGY Clinical evidence supports the use of barley (> 3 g β-glucans per day) in reducing serum (LDL) cholesterol and controlling diabetes.

TOXICOLOGY Non-toxic (edible). Hordenine has a low toxicity: LD_{50} = 299 mg/kg (mouse, i.p.).

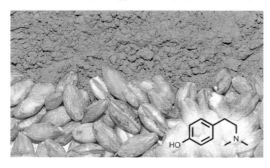

Hordeum vulgare L. (Poaceae); *mai ya* (Chinese); *orge* (French); *Gerste* (German); *orzo* (Italian); *cebada* (Spanish)

Hovenia dulcis
Japanese raisin tree

CLASSIFICATION TM: Asia (China). DS: alcohol antagonist.

USES & PROPERTIES The edible fruit stalks have been used in Chinese traditional medicine as a hangover cure. Extracts and health drinks have become popular to counteract the effects of alcoholism and damage to the liver.

ORIGIN East Asia (China, Japan, Korea and the Himalayas). It is cultivated as an ornamental tree and source of commercial product.

BOTANY Deciduous tree (to 10 m); leaves simple, glossy; flowers small, white; fruit small, dry, inedible, fruit stalks branched, fleshy, juicy, edible.

CHEMISTRY Stalks contain flavonoids: **ampelopsin** (a flavanol) is a major compound. Hodulcine 1 (a triterpenoid glycoside in leaves) inhibits sweet taste perception (less active than gymnemic acid).

PHARMACOLOGY Animal studies have shown that ampelopsin has hepatoprotective activity and counteracts the effects of alcohol.

TOXICOLOGY The fruit stalks are edible and no side effects have been reported.

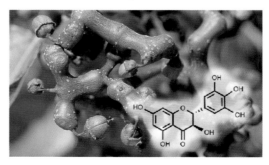

Hovenia dulcis Thunb. (Rhamnaceae); *kouai tsao* (Chinese); *raisin de chine* (French); *Japanischer Rosinenbaum* (German); *ovenia dolce* (Italian); *kemponashi* (Japanese)

Humulus lupulus
hop plant

CLASSIFICATION Sedative (III). TM: Europe. Comm.E+, ESCOP 4, WHO 3, PhEur8, HMPC.
USES & PROPERTIES Hops are the dried, cone-like female flower clusters (*Lupuli strobulus*), while hops grains are small resinous glands obtained by sieving (*Lupuli glandula*). Both are used in traditional medicine as bitter tonics and diuretics. In recent years, hops has become popular as an ingredient of natural sedative mixtures and sleep-promoting teas.
ORIGIN Asia, Europe and North America. Widely cultivated for hops (used mainly in beer brewing).
BOTANY Perennial creeper (vine) (to 10 m); leaves deeply lobed; female flowers in cone-like clusters.
CHEMISTRY A C5 alcohol (**2-methyl-3-buten-2-ol**) forms in hops during storage. The bitter taste and antibacterial activity of hops are due to lupulone and humulone (phloroglucinol derivatives).
PHARMACOLOGY Experiments showed that 2-methyl-3-buten-2-ol has strong sedative activity. It may be formed from lupulone after oral intake.
TOXICOLOGY No side effects are known.

Humulus lupulus L. (Cannabaceae); *houblon* (French); *Hopfen* (German); *luppolo* (Italian); *lupulo* (Spanish)

Hydrastis canadensis
goldenseal

CLASSIFICATION Neurotoxin, mind-altering (Ib–II). TM: North America, Europe. PhEur8, Homeopathy.
USES & PROPERTIES The dried rhizome and root (*Hydrastis rhizoma*), taken as an infusion of 0.5–1 g three times per day, are traditionally used to stop bleeding and diarrhoea. It is a general medicine and bitter tonic, digestive stimulant, mild laxative and antihaemorrhagic. Extracts are used in washes and gargles for mouth infections and in eye drops.
ORIGIN North America. Wild-harvesting is not sustainable but alternative sources can be used.
BOTANY Deciduous perennial herb; leaves lobed; flowers solitary, white; fruits fleshy, red, inedible.
CHEMISTRY Isoquinoline alkaloids: mainly **hydrastine** (1.5–4%) and berberine (to 6%).
PHARMACOLOGY Antibacterial activity is due to the alkaloids. They affect many different receptors, enzymes and DNA (berberine).
TOXICOLOGY The alkaloids are toxic is large doses and may cause vomiting, digestive disturbances, uterus contraction, hallucinations and delirium.

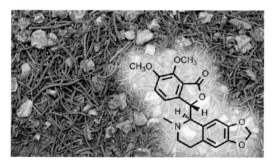

Hydrastis canadensis L. (Ranunculaceae); *hydrastis* (French); *Kanadische Gelbwurzel* (German); *sigillo d'oro* (Italian)

Hyoscyamus niger
henbane

CLASSIFICATION Neurotoxin, mind-altering, extremely hazardous (Ia). TM: Europe. Pharm., Comm.E+. MM: alkaloids, clinical studies+.

USES & PROPERTIES Mainly the leaves (*Hyoscyami folium*) but also the seeds (*Hyoscyami semen*) or roots (*Hyoscyami radix*) are traditionally used to treat nervous disorders, pain and toothache. It was once smoked to treat asthma, applied as narcotic in surgery (and as hallucinogen in witchcraft). The daily dose of standardised henbane powder (0.5–0.7 mg/g alkaloid) is 3 g, with a maximum single dose of 1 g. It is also a mind-altering drug.

ORIGIN Europe and Asia (a weed of disturbed sites, naturalised in North America and Australia).

BOTANY Annual or biennial herb (to 0.5 m); leaves lobed, hairy; flowers pale yellow, veined purple.

CHEMISTRY Tropane alkaloids: **hyoscyamine** is the main compound (or atropine, its racemate) and scopolamine.

PHARMACOLOGY The alkaloids have antispasmodic, sedative and narcotic effects.

TOXICOLOGY Lethal dose in children: 15 seeds.

Hyoscyamus niger L. (Solanaceae); *jusquiame noire* (French); *Bilsenkraut* (German); *giusquiamo nero* (Italian); *veleño negro* (Spanish)

Hypericum perforatum
St John's wort

CLASSIFICATION TM: Europe. Comm.E+, ESCOP 1, WHO 2, PhEur8, HMPC, clinical studies+.

USES & PROPERTIES Dried flowering tops (St. John's wort – *Hyperici herba*) have a long tradition as medicinal product but are used in modern times mainly to treat mild depression, anxiety and mood disturbances. A bright red oil (infusion in vegetable oil) is used for treating wounds and burns.

ORIGIN Europe and Asia. Naturalised in North America, Australia and South Africa.

BOTANY Perennial shrublet (to 0.6 m); leaves small, gland-dotted; flowers bright yellow.

CHEMISTRY The main active compounds are hypericin (a dianthrone), hyperforin (a phloroglucinol derivative) and flavonoids.

PHARMACOLOGY The drug functions as a reuptake inhibitor, hence the antidepressant activity (shown in clinical studies). It has antimicrobial and antiviral activities.

TOXICOLOGY Hypericin is phototoxic and may cause severe blistering of the skin at high doses. It interacts with pharmaceutical medicines.

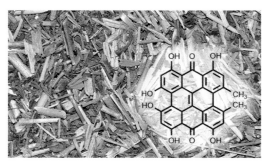

Hypericum perforatum L. (Hypericaceae); *millepertuis perforé* (French); *Echtes Johanniskraut, Tüpfel-Johanniskraut* (German); *iperico, erba di San Giovanni* (Italian)

Krameria lappacea
rhatany • Peruvian rhatany

CLASSIFICATION TM: Europe, Central and South America. Pharm., Comm.E+, ESCOP, PhEur8.

USES & PROPERTIES The dried root (*Ratanhiae radix*) is traditionally used to treat diarrhoea and infections of the mouth and throat. A decoction is made of 1.5–2 g of the root or commercial tinctures can be used (5–10 drops in a glass of water, 3 × per day). Undiluted tincture can be applied directly to the tongue, throat or gums. Rhatany root has been used as a venotonic to alleviate the symptoms of capillary fragility and haemorrhoids, and to treat burns and bleeding wounds.

ORIGIN South America (Bolivia, Ecuador, Peru).

BOTANY Shrub (to 1 m); leaves small, silky-hairy; flowers purplish red, with four petals.

CHEMISTRY Condensed tannins (15%) and neolignans (ratanhia phenols), with **ratanhiaphenol II** and (+)-conocarpan as main compounds.

PHARMACOLOGY The tannins are antimicrobial, astringent and antidiarrhoeal. Topical anti-inflammatory activity is due to the lignans.

TOXICOLOGY Low doses have no side effects.

Krameria lappacea (Domb.) Burd. & Simp. [= *K. triandra* Ruíz & Pavón] (Krameriaceae); *ratanhia* (French); *Ratanhia* (German); *ratania* (Italian); *ratania* (Spanish)

Laburnum anagyroides
golden chain • golden rain

CLASSIFICATION Neurotoxin, highly hazardous (Ib). TM: Europe.

USES & PROPERTIES The tree is a popular ornamental plant, commonly grown in gardens and parks. Young children are attracted to the fruits, which superficially resemble those of beans and peas. As a result, *Laburnum* is a leading cause of intoxications reported to poison centres in Europe.

ORIGIN Central and southeast Europe.

BOTANY Long-lived shrub or small tree (usually 5–10 m); leaves trifoliate; flowers yellow, showy, in long pendulous clusters; fruit an oblong, flat, 4–6-seeded pod; seeds black.

CHEMISTRY **Cytisine** and other quinolizidine alkaloids (3% in seeds, 0.3% in leaves).

PHARMACOLOGY Symptoms of cytisine poisoning set in after 15–60 minutes and include nausea, vomiting, cold sweat, confusion and spasms. Cytisine is an AChR agonist and affects the nervous system.

TOXICOLOGY Three to four pods (15 to 20 seeds) can cause serious poisoning in children.

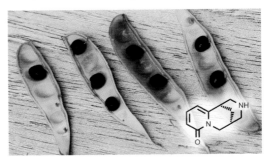

Laburnum anagyroides (L.) Medikus (Fabaceae); *aubour, cytise faux ébénier* (French); *Goldregen* (German); *avorniello* (Italian)

Lamium album
white dead nettle

CLASSIFICATION TM: Europe. Pharm., Comm. E+ (flowers only).
USES & PROPERTIES The dried flowering tops (white dead nettle herb – *Lamii albi herba*) or the dried petals (white dead nettle flower – *Lamii albi flos*) are traditionally used as an expectorant to dissolve phlegm in catarrh of the upper respiratory tract, as well as for gastrointestinal complaints. The daily dose is 3 g. Other uses that have been recorded include menopausal and urogenital disorders, and leucorrhoea. Poultices have been applied to bruises, swellings and varicose veins.
ORIGIN Europe and Asia.
BOTANY Perennial herb (to 0.5 m), leaves opposite, resembling nettle; flowers 2-lipped, white.
CHEMISTRY Iridoid glucosides (mainly **lamalbide**); also tannins, triterpene saponins and acids.
PHARMACOLOGY Anti-inflammatory, weak diuretic and antiseptic activities are ascribed to the iridoids, tannins and saponins.
TOXICOLOGY No contraindications or side effects are known.

Lamium album L. (Lamiaceae); *ortie blanche* (French); *Weiße Taubnessel* (German); *ortica bianca* (Italian); *ortiga blanca* (Spanish)

Lathyrus sativus
grass pea • chickling vetch

CLASSIFICATION Cell toxin (II).
USES & PROPERTIES Grass pea seeds are apparently not toxic when consumed in combination with other staple foods. However, during times of famine, when people have nothing else to eat, it causes lathyrism (paralysis of the lower limbs, with the knees often locked in a bent position). Outbreaks were recorded in France in 1700–1701 and 1856. Grass pea flour was banned in Germany in 1671.
ORIGIN Eastern Mediterranean. Still an important food source in Ethiopia, Pakistan, India and Bangladesh despite efforts to discourage its use. It is highly resistant to drought and will provide at least some harvest when all other crops fail.
BOTANY Annual herb; stems winged, with climbing tendrils; leaves trifoliate; flowers blue; fruit a 3–5-seeded pod; seeds mottled.
CHEMISTRY Seeds contain a toxic non-protein amino acid, **oxalyldiaminopropionic acid** (ODAP).
PHARMACOLOGY ODAP is an analogue of the neurotransmitter glutamate and causes lathyrism.
TOXICOLOGY Cooking removes 90% of the toxin.

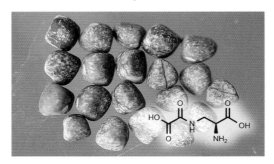

Lathyrus sativus L. (Fabaceae); *ou zhou xiang wan dou* (Chinese); *gesse blanche* (French); *Saatplatterbse* (German); *khesari* (Hindi); *cicerchia* (Italian); *almorta* (Spanish)

Lavandula angustifolia
lavender

CLASSIFICATION TM: Europe. Comm.E+, ESCOP, PhEur8, HMPC.

USES & PROPERTIES Flowers, stripped from their stalks (*Lavandulae flos*) are used as mild sedative to treat nervous conditions, excitement and sleep disturbances. It is taken for digestive ailments (as cholagogue, spasmolytic, carminative and diuretic). It is a popular ingredient of calming teas, sedative preparations, tonics and cholagogues. Lavender oil (*Lavandulae aetheroleum*) is used on the skin for antiseptic, soothing and wound-healing effects; in aromatherapy, for relief of nervous tension, emotional upsets, headache and migraine.

ORIGIN Western Mediterranean. Widely cultivated (ornamental, source of essential oil).

BOTANY Aromatic shrublet (to 1 m); leaves silvery; flowers purple-blue, in elongated spikes.

CHEMISTRY Essential oil, dominated by **linalyl acetate** and linalool.

PHARMACOLOGY Sedative, spasmolytic and mild antimicrobial activities (due to terpenoids).

TOXICOLOGY The oil is toxic when ingested.

Lavandula angustifolia Mill. [= *L. officinalis* Chaix, *L. vera* DC.] (Lamiaceae); *lavande* (French); *Echter Lavendel* (German); *lavanda* (Italian); *lavanda* (Spanish)

Lawsonia inermis
henna

CLASSIFICATION TM: Asia (India).

USES & PROPERTIES The powdered leaf (*Hennae folium*) is used in Ayurvedic medicine to treat burns, wounds and other skin ailments and may be taken orally as antidiarrhoeal, anti-epileptic and abortifacient. It is used to colour hair and nails, and for traditional body art (intricate decorations on hands and feet).

ORIGIN South Asia, Middle East and Mediterranean region. It is widely cultivated.

BOTANY Evergreen shrub (5 m); leaves opposite; flowers white; fruit a small, many-seeded capsule.

CHEMISTRY The active compound is **lawsone**, a 2-hydroxy-1,4-naphthoquinone that occurs in the intact plant as glycoside (1% in dry leaf powder) but also as free aglycone. Also present are tannins (10%), coumarins, flavonoids, phenolic acids, sterols and xanthones.

PHARMACOLOGY Astringent and antiseptic activities are due to tannins and lawsone.

TOXICOLOGY There are some indications that lawsone may be mutagenic.

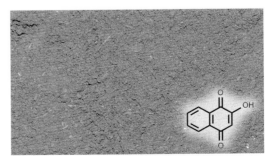

Lawsonia inermis L. (Lythraceae); *henné* (French); *Hennastrauch* (German); *hennè* (Italian)

Leonurus cardiaca
motherwort

CLASSIFICATION Mild narcotic and sedative (III). TM: Europe. Pharm., Comm.E+., WHO 5, PhEur8, HMPC.
USES & PROPERTIES The aboveground parts (*Leonuri herba*) are traditionally used to treat nervous heart conditions (reflected in *cardiaca*) and anxiety during childbirth (hence "motherwort"). It is also used as spasmolytic and hypotensive, and in supportive treatment of an overactive thyroid.
ORIGIN Central Europe, northern Europe and Asia. The herb is commonly grown in gardens.
BOTANY Erect perennial herb (to 1.5 m); leaves toothed, drooping; flowers pink, hairy. Chinese- (*L. heterophyllus*) and Siberian motherwort (*L. sibirica*) have been used in the same way.
CHEMISTRY Iridoid glucosides (**ajugol**, ajugoside), alkaloids (stachydrine, leonurine), diterpenes (leocardin, marrubiaside), flavonoids, tannins and phenolic acids.
PHARMACOLOGY Demonstrated cardiotonic and uterotonic activities are not yet explained.
TOXICOLOGY Considered safe to use at 4.5 g/day.

Leonurus cardiaca L. (Lamiaceae); *agripaume* (French); *Herzgespann* (German); *cardiaco* (Italian); *agripalma* (Spanish)

Lepidium sativum
garden cress • cress

CLASSIFICATION TM: Africa, Arabia, Asia, Europe, North America. DS: functional food.
USES & PROPERTIES Seeds are traditionally used in African, Arabian and Asian medicine for many ailments, including asthma, coughs and bronchitis (also as antiseptic, diuretic, expectorant, aphrodisiac, stomachic and laxative). In Europe and North America, the herb is used as food (formerly as diuretic, anthelmintic and antiscorbutic).
ORIGIN Uncertain: western Asia (Iran) or Africa (Ethiopia). Used in ancient Egypt as food. It is cultivated in Europe, Asia and North America.
BOTANY Annual herb (to 0.6 m); leaves compound; flowers white; fruit a bilocular capsule.
CHEMISTRY **Glucotropaeolin** and other glucosinolates; lepidine and other imidazole alkaloids; flavonoids, mucilages, fatty acids, vitamin C.
PHARMACOLOGY Diuretic, airway relaxant, antihypertensive, anti-inflammatory, antirheumatic, analgesic, antispasmodic, aphrodisiac, hypoglycaemic, antibacterial and many other activities.
TOXICOLOGY Non-toxic (edible).

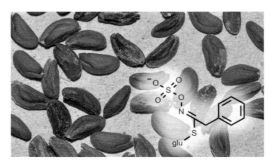

Lepidium sativum L. (Brassicaceae); *jia du xing cai* (Chinese); *cresson alénois* (French); *Gartenkresse* (German); *crescione* (Italian); *mastruco* (Portuguese); *berro de huerta* (Spanish)

Lessertia frutescens
sutherlandia • cancer bush

CLASSIFICATION TM: Africa. AHP.

USES & PROPERTIES Dried twigs and leaves (1–3 g/day) are a traditional bitter tonic and adaptogen to treat many ailments. The main uses are against stress, diabetes and cancer (as treatment and prophylaxis) but also colds, influenza, cough, asthma, bronchitis, fever, indigestion, heartburn, poor appetite, gastritis, peptic ulcers, diarrhoea, liver conditions, rheumatism, urinary tract infections, tuberculosis, and topically for skin disorders.

ORIGIN Southern Africa. Cultivated as a garden plant and also commercially for medicinal use.

BOTANY Perennial shrub (to 2 m); leaves pinnate; flowers red; fruit an inflated, many-seeded pod.

CHEMISTRY Triterpenoid saponins (**SU1**); NPAAs, *L*-canavanine; flavonoids; pinitol.

PHARMACOLOGY The triterpenoids have bitter tonic, anticancer and possible corticomimetic effects. Canavanine has anticancer and antiviral activity. Pinitol is an antidiabetic and is potentially useful to treat wasting in cancer and AIDS.

TOXICOLOGY No serious side effects are known.

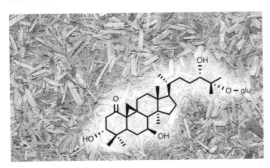

Lessertia frutescens (L.) P.Goldblatt & J.C.Manning [= *Sutherlandia frutescens* (L.) R.Br.] (Fabaceae); *sutherlandia* (French); *Sutherlandia* (German); *sutherlandia* (Italian)

Levisticum officinale
lovage

CLASSIFICATION TM: Europe. Pharm., Comm. E+, PhEur8, HMPC.

USES & PROPERTIES The dried rhizome and roots (*Levistici radix*) are used as diuretic to treat oedema, inflammation of the lower urinary tract and to prevent kidney gravel. The dose is 1.5–3 g, taken 3 × per day as tea (or half an hour before meals, when used as stomachic, carminative, expectorant and emmenagogue). Lovage root is an ingredient of commercial urological and cardiotonic preparations. Fruits and leaves are also used.

ORIGIN Eastern Mediterranean. Cultivated (and naturalised) in Europe and North America as a popular spice and ingredient of bitter liqueurs.

BOTANY Perennial herb (to 2 m); leaves compound, aromatic; flowers small; fruit a schizocarp.

CHEMISTRY Essential oil (1%) with alkylphthalides (e.g. 3-butylphthalide, **ligustilide**, sedanolide).

PHARMACOLOGY The phthalides are antispasmodic, carminative and sedative. They also increase the flow of saliva and gastric juices.

TOXICOLOGY Edible (the tea has no side effects).

Levisticum officinale Koch [= *Ligusticum levisticum* L.] (Apiaceae); *livèche* (French); *Liebstöckel, Maggikraut* (German); *levistico* (Italian); *ligustico* (Spanish)

Linum usitatissimum
flax • linseed

CLASSIFICATION Cell toxin (II). TM: Europe. Pharm., Comm. E+, ESCOP 1, PhEur8, HMPC, clinical studies+. DS: functional food.
USES & PROPERTIES Ripe, dried seeds (linseed; *Lini semen*) are used for ongoing constipation, gastritis, enteritis, diverticulitis, irritable colon, ulcerative colitis and diarrhoea. Whole or cracked seeds (1 tablespoon in 150 ml water, 3 × per day).
ORIGIN Mediterranean and western Europe.
BOTANY Erect annual herb (to 1 m); leaves small; flowers pale blue; fruit a few-seeded capsule.
CHEMISTRY Oil (35–45%): linoleic and α-linolenic acids; mucilage (6–10%), in the outer layer of the seed coat; cyanogenic glucosides (1%), mainly **linamarin**; lignans: secoisolariciresinol (SDG); proteins (25%); fibre (25%).
PHARMACOLOGY Bulk laxative. Clinical evidence for benefits in high cholesterol and diabetes. SDG: possibly anticarcinogenic and phytoestrogenic.
TOXICOLOGY The seeds are edible and safe to consume. Linamarin and other cyanogenic glucosides should not pose a health risk in therapeutic doses.

Linum usitatissimum L. (Linaceae); *lin* (French); *Lein, Flachs* (German); *lino* (Italian); *lino* (Spanish)

Lobelia inflata
Indian tobacco • asthma weed

CLASSIFICATION Neurotoxin, mind-altering (Ib).TM: Europe, North America. Pharm. MM: alkaloid (lobeline).
USES & PROPERTIES Stems and leaves (*Lobeliae herba*) are used to treat asthma, bronchitis and whooping cough. An infusion or decoction of 0.2–0.6 g of the herb is taken 3 × per day. Extracts and tinctures are also taken (sometimes as commercial mixtures). Isolated lobeline is an ingredient of oral preparations to treat nicotine withdrawal. Injection of extracts to resuscitate babies with asphyxia and apnoea is no longer considered safe.
ORIGIN North America.
BOTANY Annual herb (to 0.5 m); leaves sessile; flowers pale blue, calyx inflated; fruits a capsule.
CHEMISTRY Piperidine alkaloids (0.5%): **lobeline** is the main compound in this and other species.
PHARMACOLOGY Lobeline activates nicotinic acetylcholine receptors and acts as a respiratory stimulant and bronchodilator.
TOXICOLOGY Safe in low doses only. The alkaloids are very poisonous when used in large amounts.

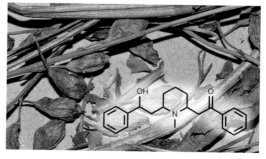

Lobelia inflata L. (Campanulaceae); *lobélie enflée* (French); *Aufgeblasene Lobelie, Indianertabak* (German); *lobelia* (Italian)

Lophophora williamsii
mescal • peyote

CLASSIFICATION Neurotoxin, hallucinogen, mind-altering (Ib–II). TM: Central America.

USES & PROPERTIES About 10 to 12 dried tops of the stems ("mescal buttons") are chewed to produce intoxicating and hallucinogenic effects. Drug users apply pure alkaloid by injection. The drug has a long history of use by the Aztecs (as *peyotl*) and the isolated mescaline by artists and authors.

ORIGIN Central America.

BOTANY Globular cactus (to 100 mm in diameter), bluish green; thorns absent; flowers pink.

CHEMISTRY Phenylpropylamines: **mescaline** is the main alkaloid (4.5–7%). Modern designer drugs such as Ecstasy resemble mescaline in chemical structure and in psychedelic effects.

PHARMACOLOGY Mescaline is a serotonin agonist, a stimulant and hallucinogen. Doses above 400 mg cause paralysis of the central nervous system (with low blood pressure, weakness, liver damage and vomiting).

TOXICOLOGY Mescaline has a relatively low toxicity. LD_{50} = 212 mg/kg (mouse, i.p.).

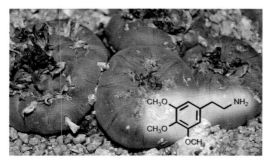

Lophophora williamsii (Lem. ex Salm-Dyck) J.M.Coult. (Cactaceae); *peyote, peyoti* (French); *Meskalkaktus, Peyotekaktus* (German)

Lycium chinense
Chinese wolfberry • goji berry

CLASSIFICATION TM: Asia (China). DS: functional food.

USES & PROPERTIES Fresh or more often dried berries of both *Lycium chinense* and *L. barbarum* (*gou qi zi* – *Lycii fructus*) are a traditional tonic in China, used to improve eyesight, protect the liver and treat wasting diseases. Dried roots (*Lycii radicis cortex*): fever and high blood pressure.

ORIGIN Eastern Asia (*L. chinense*) and China (Ningxia Province, *L. barbatum*). Both are naturalised in North America (*L. barbarum* in Europe).

BOTANY Woody shrub (1 m); stems spiny; leaves lanceolate; flowers purple, tube 1.5 mm long; fruit a drupe, 10 mm long (*L. barbarum*: leaves ovate, flower tube 2.5 mm, fruit to 20 mm long).

CHEMISTRY Amino acids (to 5%), proteoglycan polysaccharides (30% of fruit pulp), antioxidant carotenoids (high levels of **zeaxanthin** and β-carotene), triterpenes and saponins.

PHARMACOLOGY The health benefits are linked to the nutrients, polysaccharides and carotenoids.

TOXICOLOGY Fruits are edible (non-toxic).

Lycium chinense Mill. (Solanaceae); *lyciet* (French); *Chinesischer Bocksdorn* (German); *licio* (Italian)

Lycopersicon esculentum
tomato

CLASSIFICATION Cytotoxin and neurotoxin (II) (green fruits). DS: lycopene.
USES & PROPERTIES The red pigment in ripe fruits (lycopene) is an important dietary supplement. Cooked and processed tomato is second only to gac fruit (*Momordica cochinchinensis*) as a source. The latter is sold as soft capsules and juice blends with high levels of lycopene and β-carotene. Lycopene is fat-soluble.
ORIGIN South and Central America. Early domestication may have occurred in Mexico.
BOTANY Short-lived perennial (grown as annual); leaves aromatic; flowers yellow; fruit a large berry.
CHEMISTRY Carotenoid pigments: mainly **lycopene** (ca. 40 μg/g in raw tomatoes, 60–130 μg/g in tomato sauce and ketchup).
PHARMACOLOGY Lycopene is a powerful antioxidant but not a vitamin A precursor. Lycopene is claimed to reduce the risk of prostate cancer.
TOXICOLOGY Lycopene is an approved natural food colouring (non-toxic). Green tomatoes may contain toxic levels of steroidal alkaloids.

Lycopersicon esculentum Mill. or ***Solanum lycopersicum*** L. (Solanaceae); *fan qie* (Chinese); *tomate* (French); *Tomate* (German); *pomodoro* (Italian); *tomate* (Portuguese, Spanish)

Lycopodium clavatum
common clubmoss

CLASSIFICATION Neurotoxin, mind-altering (II). TM: Europe, Asia.
USES & PROPERTIES The whole herb (*Lycopodii herba*) is traditionally used as a diuretic for the treatment of inflammations of the bladder and urinary tract, as well as menstrual ailments. The daily dose is 4.5 g. It is used as a tonic in Chinese Traditional Medicine. Pure alkaloids such as huperzine A show potential for treating Alzheimer's disease.
ORIGIN Arctic region (Eurasia, North America).
BOTANY A clubmoss (Lycopodiales, Pteridophyta) with creeping and branching stems (to 1 m); leaves small, linear, spirally arranged.
CHEMISTRY Quinolizidine alkaloids: **lycopodine** (the main compound), with dehydrolycopodine and others; also triterpenes, flavonoids and acids.
PHARMACOLOGY The alkaloids are sedative and antispasmodic. Huperzine A inhibits acetylcholinesterase but clinical use for dementia is still experimental.
TOXICOLOGY High doses of alkaloids can cause nausea, dizziness, staggering and coma.

Lycopodium clavatum L. [= *L. vulgare*] (Lycopodiaceae); *lycopode en massue, jalousie* (French); *Keulen-Bärlapp* (German); *licopodio clavato* (Italian)

Lycopus europaeus
bugleweed • gipsywort

CLASSIFICATION TM: Europe, North America. Comm. E+, clinical studies+.

USES & PROPERTIES The dried aboveground parts, harvested from flowering plants (*Lycopi herba*), are traditionally used to treat an overactive thyroid gland, heart palpitations and mastodynia. Daily dose (adapted to individual need): 1–2 g, as infusion. Extracts are ingredients of commercial preparations. In Chinese medicine, *L. lucidus* (top right) is used to treat gynaecological ailments.

ORIGIN Europe and Asia.

BOTANY Erect perennial herb (to 0.5 m); leaves opposite, toothed; flowers small, subsessile.

CHEMISTRY Phenolic acids (hydroxycinnamic acid derivates): **lithospermic acid** and rosmarinic acid; flavonoids; terpenes and essential oil.

PHARMACOLOGY Depsides of the hydrocinnamic acids are linked to the antithyrotropic activity, which is supported by clinical evidence.

TOXICOLOGY High doses and continued use may lead to thyroid hypertrophy. Contraindicated for persons with thyroid insufficiency.

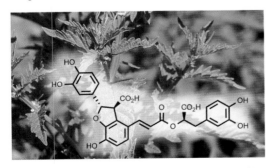

Lycopus europaeus L. (Lamiaceae); *pied-de-loup* (French); *Gemeiner Wolfstrapp* (German); *marrubio d'acqua* (Italian); *manta de lobo* (Spanish)

Mahonia aquifolium
Oregon grape

CLASSIFICATION Cell toxin and neurotoxin (II). TM: Europe, North America.

USES & PROPERTIES The root and root bark are used as bitter tonic and general medicine for digestive ailments, liver complaints and psoriasis. It is also used as cholagogue, diuretic and laxative. Similar to *Berberis* species and used in much the same way. Extracts are active ingredients of commercial preparations for treating psoriasis. The efficacy is sometimes questioned, despite clinical studies that show benefits.

ORIGIN North America. A popular and attractive garden subject, found in many parts of the world.

BOTANY Evergreen shrub; leaves compound, spiny; flowers yellow; fruit a small purple drupe.

CHEMISTRY Isoquinoline alkaloids: mainly berberine. Also **5'-methoxyhydnocarpin** (shown to potentiate the antimicrobial action of berberine).

PHARMACOLOGY The alkaloids have antimicrobial, cytotoxic, hypotensive and cholekinetic activities; possible benefits in treating psoriasis.

TOXICOLOGY Toxic dose of berberine: >0.5 g.

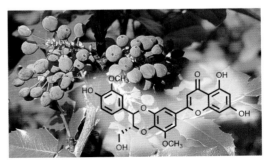

Mahonia aquifolium (Pursh) Nutt. or ***Berberis aquifolium*** Pursh (Berberidaceae); *mahonie a feuilles de houx* (French); *Gewöhnliche Mahonie* (German); *maonia commune* (Italian)

Malva sylvestris
mallow • common mallow

CLASSIFICATION TM: Europe. Pharm., Comm. E+, ESCOP, PhEur8.
USES & PROPERTIES The dried leaves (*Malvae folium*) or flowers (*Malvae flos*) are traditionally used against colds, catarrh, inflammation of the mouth and throat, as well as gastroenteritis. Both are used as poultice to treat wounds. The flowers (of subsp. *mauritiana*) are a traditional diuretic in folk medicine and are also used as a natural dye to colour teas and foodstuffs.
ORIGIN Europe and Asia. Naturalised in many parts of the world.
BOTANY Biennial herb (to 1 m); leaves deeply lobed; flowers pink (purple in subsp. *mauritiana*).
CHEMISTRY Mucilage (8–10%); flavonoids sulfates in the leaves; anthocyanins – mainly **malvin** (malvidin 3,5-diglucoside) in the flowers (petals).
PHARMACOLOGY The mucilage is demulcent, forming a protective layer over inflamed mucosa. Anthocyanins provide colour but are also mildly astringent, antioxidant and anti-inflammatory.
TOXICOLOGY No side effects are known.

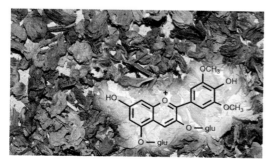

Malva sylvestris L. (Malvaceae); *mauve sauvage* (French); Wilde Malve, Große Käsepappel (German); *malva riondela* (Italian); *malva* (Spanish)

Mandragora officinarum
mandrake

CLASSIFICATION Neurotoxin, hallucinogen, mind-altering (Ia). TM: Europe.
USES & PROPERTIES Mandrake has been used since ancient times as anthelmintic, analgesic, hypnotic, aphrodisiac and narcotic for surgery. It is mentioned in the Ebers Papyrus (1500 BC) and in the famous *Materia Medica* written by Dioscorides between 50 and 70 BC. It is associated with Greek mythology and medieval witchcraft and sorcery.
ORIGIN Southern Europe.
BOTANY Small perennial herb; taproot up to 0.6 m long (according to tradition it resembles a human body); flowers purple; fruit fleshy, many-seeded.
CHEMISTRY Tropane alkaloids: mainly **scopolamine** (hyoscine), with hyoscyamine and others.
PHARMACOLOGY Scopolamine is an antagonist at the muscarinic acetylcholine receptor and acts as a parasympatholytic. It is depressant and sedative at low doses but euphoric and hallucinogenic at high doses (causes a feeling of flying).
TOXICOLOGY The lethal dose of scopolamine is 100 mg in adults and 2–10 mg in children.

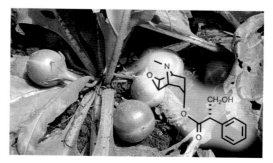

Mandragora officinarum L. (Solanaceae); *mandragore* (French); Alraune (German); *mandragora primaverile* (Italian)

Marrubium vulgare
white horehound

CLASSIFICATION TM: Europe. Pharm., Comm. E+, PhEur8, HMPC.

USES & PROPERTIES Fresh or dried aboveground parts (*Marrubii herba*) are traditionally used for menstrual disorders and inflammations of the skin and mucosa. It is nowadays used for digestive ailments (indigestion, flatulence, biliary complaints and lack of appetite) and dry coughs associated with chronic bronchitis. The daily dose is 4.5 g, taken as tea. Commercial expectorants and digestive medicines sometimes contain horehound.

ORIGIN Southern Europe and Asia. It has become a weed in many warm parts of the world.

BOTANY Perennial herb (to 0.5 m); stems square; leaves prominently veined; flowers small, two-lipped, white, in rounded clusters.

CHEMISTRY Diterpene lactones (**marrubiin**, 1%); also flavonoids, tannins, acids and essential oil.

PHARMACOLOGY Marrubiin has choleretic and expectorant activities (it stimulates secretions).

TOXICOLOGY High doses can be cardioactive and uterine stimulant (dosage should be controlled).

Marrubium vulgare L. (Lamiaceae); *marrube blanc* (French); *Gemeiner Andorn* (German); *marrobio* (Italian); *marrubio* (Spanish)

Matricaria chamomilla
chamomile • German chamomile

CLASSIFICATION TM: Europe. Pharm., Comm. E+, ESCOP 6, WHO 1, PhEur8, HMPC.

USES & PROPERTIES Dried flower heads (*Matricariae flos*): flatulent nervous dyspepsia in adults and children, as well as diarrhoea, gastritis, mild anxiety and travel sickness. Essential oil (*Matricariae aetheroleum*) or liquid extracts: applied to wounds and inflammations of the skin and mucous membranes. Oil can be inhaled: nasal catarrh and infections of the respiratory tract.

ORIGIN Eastern Europe and the Near East. It is cultivated on a large commercial scale.

BOTANY Annual herb; leaves compound, feathery; flower heads yellow, ray florets white, the disc hollow (solid in Roman chamomile – *Chamaemelum*).

CHEMISTRY Essential oil (deep blue colour), with sesquiterpenes: **α-bisabolol** (50%) and chamazulene (15%). Also present are flavonoids, coumarins, polyacetylenes and polysaccharides.

PHARMACOLOGY Sesquiterpenoids: anti-inflammatory, antispasmodic, carminative, antiseptic.

TOXICOLOGY Safe to use (allergies may occur).

Matricaria chamomilla L. [= *Chamomilla recutita* (L.) Rausch.; = *M. recutita* L.] (Asteraceae); *camomille* (French); *Echte Kamille* (German); *camomilla* (Italian); *camomila* (Spanish)

Medicago sativa
alfalfa • lucerne

CLASSIFICATION DS: functional food.
USES & PROPERTIES Stems and leaves (*Medicago sativae herba*) or seeds and sprouts (germinated seeds) are used as health food and general tonic to improve health, to aid in the recovery of convalescents and to help maintain cholesterol levels. No more than 5–10 g of dry herb or 40 g of seeds/sprouts can be taken 3 × per day with meals.
ORIGIN Southwestern Asia (Turkey). Alfalfa or lucerne is an important pasture and fodder crop.
BOTANY Perennial herb (to 1 m); leaves trifoliate, serrate; flowers usually blue; fruit a coiled pod.
CHEMISTRY Saponins (glycosides of **medicagenic acid**, hederagenin and soyasapogenols), steroids (β-sitosterol), isoflavonoids, coumarins, non-protein amino acids (canavanine), proteins.
PHARMACOLOGY Saponins seem to block the intestinal absorption of cholesterol.
TOXICOLOGY In the USA, alfalfa is "generally regarded as safe" but excessive quantities of seeds or sprouts in the diet can lead to systemic *lupus erythromatosus* (SLE) or reversible pancytopenia.

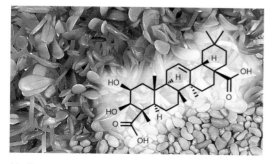

Medicago sativa L. (Fabaceae); *luzerne* (French); *Luzerne* (German); *alfa-alfa* (Italian)

Melaleuca alternifolia
tea tree

CLASSIFICATION TM: Europe, Australia. Pharm., ESCOP Suppl., WHO 2, PhEur8, HMPC, clinical studies+.
USES & PROPERTIES Essential oil (*Melaleucae aetheroleum*), extracted from stems and leaves by steam distillation, is used in creams and ointments to treat a wide range of skin ailments (including abrasions, wounds, acne, itches, infections, athlete's foot and thrush). Dilute oil can be used as gargle and mouthwash to treat inflammation of the mouth and throat, sinusitis and catarrh.
ORIGIN Australia (New South Wales). Commercially cultivated but also grown in herb gardens.
BOTANY Medium-sized tree; leaves linear; flowers without petals, white; fruit a small capsule.
CHEMISTRY Essential oil: the main monoterpenoids in high-quality oil are **terpinen-4-ol** (at least 30%), α- and γ-terpinene (40%) and 1,8-cineole (= eucalyptol; up to 15% maximum).
PHARMACOLOGY Efficacy in treating acne and fungal infections is supported by clinical studies.
TOXICOLOGY External use is safe.

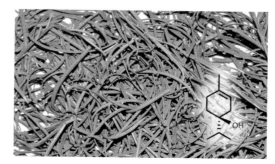

Melaleuca alternifolia Cheel. (Myrtaceae); *melaleuca* (French); *Teebaum* (German); *melaleuca* (Italian)

Melilotus officinalis
sweet clover • common melilot

CLASSIFICATION TM: Europe. Pharm., Comm. E+, ESCOP 4, PhEur8, HMPC.

USES & PROPERTIES Fresh or dried flowering tops of *M. officinalis* or *M. alba* (below) (*Meliloti herba*) are used to treat varicose veins, pruritus, acute haemorrhoids and other symptoms of venous or lymphatic insufficiency. Preparations are used for minor sleep disturbances and stomach ailments. Effective dose: the equivalent of 3–30 mg of coumarin (oral use) or 1–7.5 mg (parenteral use).

ORIGIN Europe, Asia and North Africa.

BOTANY Erect herb (to 1.2 m); leaves trifoliate, serrate; flowers yellow; fruit small, indehiscent.

CHEMISTRY The glucoside melilotoside is enzymatically converted to **coumarin**, the main ingredient of the drug; also phenolic acids, isoflavonoids and triterpenoid saponins.

PHARMACOLOGY Coumarin acts as a venotonic and decreases capillary permeability; it has anti-oedemic and anti-exudative activities.

TOXICOLOGY Anticoagulant dicoumarols may be formed; do not use while taking anticoagulants.

Melilotus officinalis (L.) Medikus (Fabaceae); *mélilot officinal* (French); *Echter Steinklee* (German); *meliloto, trifoglio cavallino* (Italian)

Melissa officinalis
lemon balm • sweet balm

CLASSIFICATION TM: Europe. Pharm., Comm. E+, ESCOP 2, WHO 2, PhEur8, HMPC.

USES & PROPERTIES The dried leaves of *Melissa officinalis* subsp. *officinalis* (*Melissae folium*) and essential oil (*Melissae aetheroleum*) are used mainly to treat nervous stomach disorders and minor sleep disturbances in adults and children. Extracts and essential oil are used topically for minor skin ailments and are included as ingredients in many commercial preparations (teas, creams, lotions).

ORIGIN Eastern Mediterranean and the Near East.

BOTANY Perennial herb (to 0.9 m); leaves opposite, prominently veined; flowers small, sessile.

CHEMISTRY Essential oil: **citronellal** (30–40%) and citral A and B (10–30%) are main compounds. Also rosmarinic acid (4%), phenolic acids, monoterpene glycosides, triterpenes and flavonoids.

PHARMACOLOGY Essential oil: sedative, carminative, spasmolytic. Ointments with essential oil and rosmarinic acid are active against *herpes simplex* and have antihormonal activity.

TOXICOLOGY No serious side effects are known.

Melissa officinalis L. (Lamiaceae); *mélisse* (French); *Zitronenmelisse* (German); *melissa, cedronella* (Italian); *melissa* (Spanish)

Mentha ×*piperita*
peppermint

CLASSIFICATION TM: Europe. Comm.E+, ESCOP 3, WHO 2, PhEur8, HMPC, clinical studies+.

USES & PROPERTIES Fresh or dried leaves (*Menthae piperitae folium*) and essential oil (*Menthae piperitae aetheroleum*) are used to treat digestive ailments and catarrhs of the respiratory tract. It is used against irritable bowel syndrome and ailments of the gall bladder and bile duct. Peppermint oil is applied topically to reduce pain and headache and as mouthwash/gargle for inflammation of the mouth and throat. It has many uses as flavourant.

ORIGIN Sterile hybrid, garden origin (England).

BOTANY Perennial herb (to 0.9 m) with rhizomes; stems purple; flowers lilac-pink. Many other species are used in traditional medicine worldwide.

CHEMISTRY Essential oil with the monoterpene **menthol** as the main active ingredient (40% or more).

PHARMACOLOGY The oil is antimicrobial, mildly analgesic, carminative, spasmolytic and choleretic. Menthol activates cold receptors to give a cooling sensation. Clinical data support the stated uses.

TOXICOLOGY The oil is unsafe to use in children.

Mentha* ×*piperita L. (Lamiaceae); *menthe poivrée* (French); *Pfefferminze* (German); *menta piperina* (Italian); *la menta* (Spanish)

Menyanthes trifoliata
bogbean

CLASSIFICATION TM: Europe. Pharm., Comm. E+, PhEur8.

USES & PROPERTIES The dried leaves (*Menyanthidis folium*) are used to stimulate appetite and to treat dyspeptic complaints. It is an ingredient of preparations used for liver and bile ailments and inflammatory conditions, mainly rheumatism. Unsweetened tea made from 0.5–1 g of dried leaves is taken half an hour before meals. The daily dose is 1.5–3 g. Extracts are used in topical anti-inflammatory products for skin disorders and rheumatism, and as bittering agent in liqueurs.

ORIGIN Northern temperate zone (North America, Europe and Asia), in lakes and marshes.

BOTANY Perennial, mat-forming herb; leaves trifoliate (like common bean); flowers white.

CHEMISTRY Bitter-tasting secoiridoid glucosides: mainly **dihydrofoliamenthin**. Coumarins (e.g. scopoletin), tannins, flavonoids and triterpenes.

PHARMACOLOGY *Amarum* (appetite stimulating) effects. Scopoletin: choleretic and cholagogic.

TOXICOLOGY There are no known side effects.

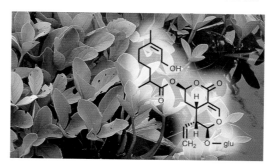

Menyanthes trifoliata L. (Menyanthaceae); *trèfle d'eau* (French); *Bitterklee, Fieberklee* (German); *trifoglio d'acqua* (Italian); *trébol acuático* (Spanish)

Mondia whitei
White's ginger • tonic root

CLASSIFICATION TM: Africa. AHP.

USES & PROPERTIES The fleshy roots are widely used across Africa as an aphrodisiac and to treat erectile dysfunction. It is used in African Traditional Medicine for many ailments, including lack of appetite, anorexia, stress and tension, nausea, indigestion, gastrointestinal disorders, constipation, diabetes, post-partum bleeding and gonorrhoea.

ORIGIN Tropical Africa (South Africa to Congo and East Africa). It is easily cultivated.

BOTANY Woody climber (vine), to 6 m; roots fleshy; leaves with frilly stipules; flowers yellow or maroon; fruit a paired follicle; seeds silky-hairy.

CHEMISTRY Simple benzaldehyde derivates: mainly 2-hydroxy-4-methoxybenzaldehyde and **isovanillin** (hence the vanilla-like smell).

PHARMACOLOGY Roots are androgenic but also show reversible antispermatogenic and antifertility effects. 2-Hydroxy-4-methoxybenzaldehyde is a specific and potent inhibitor of tyrosinase.

TOXICOLOGY Available data indicate a very low toxicity. Aldehydes are potential mutagens.

Mondia whitei (Hook.f.) Skeels (Apocynaceae); *mkombela* (Kenya); *citumbulo* (Malawi); *umondi, mundi* (Zulu); *mungurawu* (Shona)

Morinda citrifolia
noni tree • Indian mulberry

CLASSIFICATION TM: Polynesia. DS: functional food.

USES & PROPERTIES Juice from ripe fruits is used as general health drink. Traditional uses include the treatment of fever, diabetes, stomach and liver ailments, menstrual cramps and urinary tract ailments. Possible health benefits of the juice are difficult to evaluate because of the influence of commercial interests and marketing materials.

ORIGIN India, Southeast Asia and Pacific Islands (a traditional famine food and medicine).

BOTANY Evergreen shrub or tree (to 6 m); leaves glossy; flowers small, in dense heads; fruit a white compound drupe, resembling a large mulberry.

CHEMISTRY The fruit and fruit juice contain high levels of minerals and vitamins B and C. **Damnacanthal** (an anthraquinone) is present in the roots.

PHARMACOLOGY Damnacanthal is an inhibitor of tyrosine kinase and has anticancer activity. An alkaloid, xeronine, was said to be the active compound (but the structure is still unknown).

TOXICOLOGY Noni is not toxic (used as food).

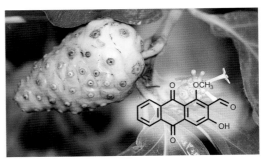

Morinda citrifolia L. (Rubiaceae); *hai ba ji* (Chinese); *nono* (French); *Indische Maulbeere, Nonibaum* (German); *bengkudu* (Malay); *nonu* (Polynesian) *mora de la India* (Spanish)

Moringa oleifera
ben tree • drumstick tree

CLASSIFICATION TM (DS): Europe, Africa, Asia.
USES & PROPERTIES Leaves, fruits and seeds are edible, while seeds, flowers and roots have been used in traditional medicine to treat a wide range of ailments (e.g. in Ayurvedic medicine, to lower blood pressure and blood glucose). Leaves are used for malnutrition relief. The seed oil, called ben or behen oil, is used as a dietary supplement and has cosmetic and industrial uses. Powdered seeds have been used to treat asthma.
ORIGIN Northwestern India; widely cultivated.
BOTANY Tree (5–15 m); leaves 2–4-pinnate; flowers creamy white; fruits to 1 m; seeds winged.
CHEMISTRY The seed oil (ben/behen oil) contains a saturated fatty acid, **behenic acid**. Antibiotic benzyl isothiocyanates are present.
PHARMACOLOGY The medicinal properties (e.g. anti-asthmatic effect) are not yet clearly explained. Benefits are derived from the high levels of nutrients (minerals, vitamins and protein) and glucosinolates. The oil has cosmetic uses.
TOXICOLOGY Leaves, fruits and seeds are edible.

Moringa oleifera Lam. (Moringaceae); *la mu* (Chinese); *ben oléifère* (French); *Meerrettichbaum* (German); *been* (Italian); *arbol de las perlas, ben* (Spanish)

Myristica fragrans
nutmeg tree

CLASSIFICATION Neurotoxin, mind-altering (II). TM: Europe, Asia. Pharm., PhEur8.
USES & PROPERTIES Dried seeds (*Myristicae semen*) and/or the dried seed arils (mace, *Myristicae arillus*) are used to treat digestive ailments and catarrh of the respiratory tract. The essential oil (*Myristicae aetheroleum*) is used topically for relief of aches and pains. Nutmeg is also an intoxicating and addictive drug.
ORIGIN Southeast Asia (Amboine Island). Cultivated in tropical regions for nutmeg and mace.
BOTANY Evergreen tree (to 20 m); flowers small, yellow; fruits (on female trees) fleshy; seed a large nut, surrounded by a bright red aril.
CHEMISTRY Essential oil: sabinene, pinene and **myristicin** with phenylpropanoids (elemicin, eugenol, isoeugenol, safrol and others).
PHARMACOLOGY Myristicin and elemicin are addictive and hallucinogenic (potentially cytotoxic, mutagenic and abortive). The oil has strong antimicrobial and anti-inflammatory activity.
TOXICOLOGY Doses exceeding 5 g are dangerous.

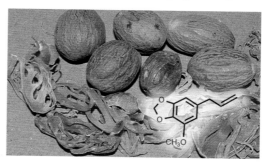

Myristica fragrans Houtt. (Myristicaceae); *noix muscade* (French); *Muskatnussbaum* (German); *noce moscata* (Italian); *nuez moscada* (Spanish)

Myroxylon balsamum
Tolu balsam tree

CLASSIFICATION TM: Central and South America. Pharm., Comm.E+, PhEur8.

USES & PROPERTIES The oleoresin obtained from damaged bark is called Tolu balsam (*Balsam tolutanum*), while smoked wood of the var. *pereirae* yields Peruvian balsam. Tolu balsam is used in cough syrups (also for asthma and whooping cough) and friar's balsam (a traditional inhalant for catarrh and colds). Peruvian balsam is applied externally to wounds, burns and bruises.

ORIGIN South and Central America.

BOTANY Tall evergreen trees (to 19 m); leaves compound; flowers white or pale blue; fruit a pod.

CHEMISTRY Balsams are oleoresins containing benzoic acid, cinnamic acid and their esters (especially **benzyl benzoate**).

PHARMACOLOGY Benzyl benzoate has vasodilating and spasmolytic effects (and is active against scabies). Benzoic acid is antibacterial; Peruvian balsam stimulates granulation in wounds.

TOXICOLOGY Benzyl benzoate has a very low oral toxicity: LD_{50} = 1700 mg/kg (rat, p.o.).

Myroxylon balsamum (L.) Harms (Fabaceae); *baumier de Tolu* (French); *Tolubalsambaum* (German); *balsamo del Tolù* (Italian)

Narcissus pseudonarcissus
wild daffodil • lent lily

CLASSIFICATION Cell toxin, highly hazardous (Ib–II). TM: Europe, Asia.

USES & PROPERTIES The bulbs have been used in traditional Roman and Japanese (Kampo) medicine as emollient to treat wounds. Daffodils are popular garden plants and poisoning may occur when the bulbs are confused with onions.

ORIGIN Western Europe. Widely naturalised and cultivated as a spring flower and cut flower.

BOTANY Perennial bulbous plant; leaves strap-shaped; flower dark or pale yellow, with a large, bell-shaped corona.

CHEMISTRY Isoquinoline alkaloids: **lycorine** and galantamine.

PHARMACOLOGY Lycorine: inhibitor of protein biosynthesis, cytotoxic, virustatic, emetic, diuretic. Galanthamine blocks cholinesterase and is used to treat Alzheimer's disease and other memory impairments (see *Galanthus nivalis*).

TOXICOLOGY Lycorine is very poisonous. LD_{50} = 41 mg/kg (dog, p.o.). Symptoms of poisoning include nausea, vomiting and slowing of heartbeat.

Narcissus pseudonarcissus L. (Amaryllidaceae); *narcisse jaune, bonhomme, chaudron* (French); *Osterglocke, Gelbe Narzisse* (German); *narciso trombone* (Italian)

Nelumbo nucifera
lotus • sacred lotus

CLASSIFICATION TM: Asia (China, India).
USES & PROPERTIES The leaves (*he ye, Folium Nelumbinis*), rhizomes (*ou jie*), seeds (*lian zi*) and seed embryos (*lian zi xin*) are all used in Chinese traditional medicine. The leaves and flowers are used to treat menorrhagia and haemorrhoids, fever, diarrhoea and nervous conditions. Seeds are a functional food, eaten to reduce nausea, indigestion, nervous conditions and insomnia.
ORIGIN Southern Asia (India and China to Australia). Widely cultivated in ponds and lakes.
BOTANY Aquatic plant; rhizome fleshy; leaves large, umbrella-shaped; flowers large, waxy, white or pink; fruit cup-shaped; seed a one-seeded nut.
CHEMISTRY Alkaloids: especially **nuciferine**. Also flavonoids in leaves and flowers.
PHARMACOLOGY Nuciferine has CNS-depressant effects (through dopamine receptor blocking) and is sedative and hypothermic. Also anti-inflammatory, hypoglycaemic and antipyretic effects.
TOXICOLOGY Nuciferine: LD_{50} = 289 mg/kg (mouse, p.o.).

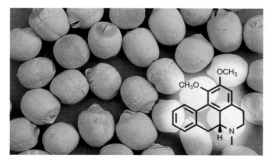

Nelumbo nucifera Gaertner [Nelumbonaceae (formerly Nymphaeaceae)]; *lian* (Chinese); *nelumbo* (French); *Lotosblume* (German); *kanwal* (Hindi); *nelumbo* (Italian); *kamala* (Sanskrit)

Nerium oleander
oleander • rose laurel

CLASSIFICATION Heart poison, extremely hazardous (II). TM: Europe. Pharm.
USES & PROPERTIES Infusions of the leaves (*Oleandri folium*) have been used in European medicine as a heart tonic to treat cardiac insufficiency. It has also been used for abortion and externally to treat skin rashes and scabies. Accidental and deliberate deaths have occurred after drinking extracts or eating only a few leaves.
ORIGIN Europe (Mediterranean region). A popular garden shrub in all warm parts of the world.
BOTANY Shrub or small tree (to 5 m); all parts exude milky latex when broken; leaves leathery; flowers in many colours; fruit oblong; seeds hairy.
CHEMISTRY Heart glycosides (cardenolides): at least 30 compounds (**oleandrin** as the main toxin).
PHARMACOLOGY Cardenolides inhibit Na^+, K^+-ATPase, leading to an increase in the force of contraction of the heart. They are also diuretic.
TOXICOLOGY The lethal oral dose for domestic animals is 0.5 mg/kg. A single leaf can cause intoxication in humans.

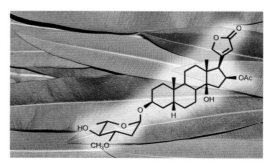

Nerium oleander L. (Apocynaceae); *laurier rose* (French); *Oleander* (German); *oleandro* (Italian); *adelfa* (Spanish)

Nicotiana glauca
tree tobacco • tobacco tree

CLASSIFICATION Neurotoxin, mutagen, highly hazardous (Ib). TM: South America.

USES & PROPERTIES The plant has been used in traditional medicine and as insecticide. Human fatalities have often been recorded, where young plants were misidentified as other plants traditionally eaten as cooked spinach (*marog*). Since a bitter taste in vegetables is often appreciated, this warning sign may not be heeded before it is too late.

ORIGIN South America (Bolivia to Argentina); an invasive weed in many parts of the world.

BOTANY Shrub or small tree (ca. 3 m); leaves fleshy, bluish-green; flowers tubular, yellow; fruit a dehiscent, many-seeded capsule.

CHEMISTRY All aboveground parts contain high levels of **anabasine** (a piperidine alkaloid) as practically the only alkaloid.

PHARMACOLOGY Anabasine is a nicotinic acetylcholine receptor antagonist that can block nerve transmission and cause death by cardiac arrest.

TOXICOLOGY Anabasine: LD_{50} = 11–16 mg/kg (mouse, p.o.).

Nicotiana glauca Graham (Solanaceae); *tabac en arbre* (French); *Baumtabak, Blaugrüner Tabak* (German); *tabacco glauco, tabacco orecchiuto* (Italian)

Nicotiana tabacum
tobacco

CLASSIFICATION Neurotoxin, mind-altering, highly hazardous (Ib). TM: South America.

USES & PROPERTIES Tobacco has been smoked since pre-Columbian times. Despite health warnings, smoking has remained popular. Tobacco may be chewed or taken as snuff. Powdered tobacco was once popular as insecticide. Pure nicotine is used in chewing gum and slow-release plasters.

ORIGIN South America; a cultigen developed from three wild species in pre-Columbian times.

BOTANY Robust annual (ca. 2 m); leaves large, soft, glandular; flower tubular, pink.

CHEMISTRY Pyridine alkaloids (to 9%): **nicotine** is the main constituent.

PHARMACOLOGY Nicotine is a potent parasympathomimetic (mAChR antagonist) that is stimulant at low doses but sedative at higher doses. It is highly addictive and may cause heart and lung ailments (especially lung carcinoma and emphysema).

TOXICOLOGY Nicotine: LD_{50} = 3 mg/kg for mice (p.o.). A dose of 0.5–1.0 mg/kg (or 50–60 mg) may be lethal for adult humans.

Nicotiana tabacum L. (Solanaceae); *tabac de Virginie* (French); *Tabak, Virginischer Tabak* (German); *tabacco Virginia* (Italian); *tabaco de Virginia* (Spanish)

Nigella sativa
black seed • kalonji

CLASSIFICATION TM (DS): Asia.
USES & PROPERTIES The ripe seeds or seed oil are important in Arabian and Islamic folk medicine. The seeds are ingested daily as general tonic and preventive medicine; also stomach ailments, colic, spasms, asthma, headache and intestinal parasites. A popular spice in the Middle East and India (often called black cumin but the latter more correctly refers to *Bunium persicum*).
ORIGIN Southern Europe, North Africa and western Asia. Cultivated since ancient times in Mesopotamia, Egypt, Arabia, Pakistan and India.
BOTANY Annual herb; leaves deeply dissected; flowers white or pale blue; fruit a many-seeded capsule; seeds black, oblong, angular.
CHEMISTRY Chemically diverse, with phytosterols, triterpene saponins, quinones, flavonols and alkaloids (nigelline). The essential oil has **thymoquinone** as main compound.
PHARMACOLOGY Immune stimulation, antispasmodic. Large doses: diuretic, stimulate lactation.
TOXICOLOGY The seeds are edible.

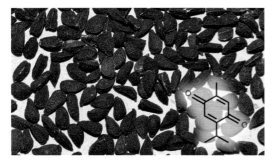

Nigella sativa L. (Ranunculaceae); *nigelle, poivrette* (French); *Schwarzkümmel* (German); *nigella* (Italian)

Ocimum tenuiflorum
holy basil • sacred basil

CLASSIFICATION TM: Asia (India). WHO 2.
USES & PROPERTIES Fresh or dried aboveground parts (*Ocimi sancti herba*) are used in Ayurvedic medicine as a general medicine and tonic (for an exceptionally wide diversity of ailments), as well as for wound healing. Infusions of the leaves or fresh leaf juice are used against cough, upper respiratory tract infections, indigestion and stress-related skin ailments. Dose: 2–4 g, three times per day.
ORIGIN Western Asia to Arabia, India, Sri Lanka, Malaysia and Australia. Widely cultivated. *Ocimum basilicum* is a popular spice.
BOTANY Short-lived perennial herb (to 1 m); leaves hairy; flowers purple, lilac or white.
CHEMISTRY Essential oil (to 2%): eugenol and methyleugenol (the main compounds), also α- and **β-caryophyllene**; tannins (4.6%) and flavonoids.
PHARMACOLOGY Tonic and wound-healing activities. Clinical data show anti-asthmatic, anti-diabetic and cholesterol-lowering activity.
TOXICOLOGY Do not use for prolonged periods. Contraindicated: pregnant women and children.

Ocimum tenuiflorum L. (Lamiaceae); *sheng luo le* (Chinese); *Basilic sacré* (French); *Indisches Basilikum* (German); *tulsi, talasi* (Gujarati); *tulasii* (Hindi)

Oenothera biennis
evening primrose

CLASSIFICATION TM: North America, Europe. WHO 2, HMPC, clinical studies+. DS: seed oil.

USES & PROPERTIES Seed oil (evening primrose oil, *Oenotherae biennis oleum*) is a dietary supplement for symptomatic relief of skin disorders: atopic eczema, pruritus and inflammation. Possible benefits: irritable bowel syndrome and menopausal, circulatory and rheumatic disorders.

ORIGIN North America (Florida to Mexico and Canada). It has become a weed in many countries.

BOTANY Biennnial or short-lived perennial herb (to 1.5 m); leaves in a basal rosette in 1st year; flowers in 2nd year, large, yellow; fruit an oblong, multi-seeded capsule (150 000 seeds per plant).

CHEMISTRY The seed oil contains high levels of **γ-linolenic acid** (gamma-linolenic acid or GLA).

PHARMACOLOGY GLA is an essential fatty acid (prostaglandin pathway). Some people are apparently unable to convert linoleic acid to GLA. Clinical studies support the use for eczema.

TOXICOLOGY The seed oil is safe to consume in recommended doses (2–3 g per day, up to 5 g).

Oenothera biennis L. (Onagraceae); *onagre bisannuelle* (French); *Gemeine Nachtkerze* (German); *enothera* (Italian)

Olea europaea
olive tree

CLASSIFICATION TM: Africa, Europe, Asia. Pharm., PhEur8, Comm.E+, HMPC.

USES & PROPERTIES Dried leaves (*Oleae folium*) are traditionally used to lower blood pressure (also as diuretic). The daily dose is about 1–2 g. Cold-pressed fruit oil (*Olivae oleum*) is taken with meals (15–30 ml) as mild laxative and cholagogue and applied externally as demulcent and emollient.

ORIGIN Mediterranean Europe, western Asia and North Africa. The tree is widely cultivated for the production of olives. African wild olive (subsp. *cuspidata*) is an alternative source of raw material.

BOTANY Evergreen tree (to 10 m); leaves opposite; flowers small, white; fruit a fleshy, oily drupe.

CHEMISTRY Bitter sesquiterpenoids: **oleuropein**, with ligustroside and oleacein. Also triterpenoids, sterols and flavonoids. Oil: oleic and linoleic acids.

PHARMACOLOGY Oleuropein lowers blood pressure by increasing coronary flow; it also has antispasmodic, antioxidant and lipid-lowering effects. Oleacein inhibits the ACE.

TOXICOLOGY Non-toxic at recommended doses.

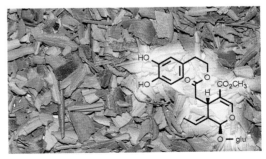

Olea europaea L. (Oleaceae); *olivier* (French); *Ölbaum* (German); *olivo, ulivo* (Italian); *olivo* (Spanish)

Ononis spinosa
spiny restharrow

CLASSIFICATION TM: Europe. Pharm., PhEur8, Comm.E+.

USES & PROPERTIES The dried roots, harvested in autumn (*Ononidis radix*) are traditionally used as a mild diuretic to treat rheumatism and gout. In modern phytotherapy it is mainly used for irrigation therapy: inflammation of the lower urinary tract and to prevent and treat kidney gravel. It is taken as an infusion of 2.5 g of root, taken several times per day (recommended daily dose: 6–12 g).

ORIGIN Europe, western Asia and North Africa.

BOTANY Shrub (to 0.8 m); stems spiny; leaves trifoliate at base, upper simple; flowers pink.

CHEMISTRY Triterpenes: mainly **α-onocerin** (onocol); sterols (mainly sitosterol); phenolic acids; isoflavones: ononine, formononetin, genistein.

PHARMACOLOGY Diuretic activity may be due to α-onocerin. Isoflavones have oestrogenic activity.

TOXICOLOGY Non-toxic (but sufficient fluids should be taken). Contraindicated for persons suffering from oedema due to cardiac or renal insufficiency.

Ononis spinosa L. (Fabaceae); *bugrane épineuse, arrête-boeuf* (French); *Dornige Hauhechel* (German); *bonaga, ononide* (Italian); *gatuña* (Spanish)

Origanum vulgare
oregano

CLASSIFICATION TM: Europe. Pharm. Comm. E+. (*O. dictamnus*: WHO 5, HMPC).

USES & PROPERTIES The dried flowering herb (*Origani vulgaris herba*) is taken as infusion (1–2 g, three times per day) to treat respiratory ailments (bronchitis, catarrh, colds, influenza) and indigestion (colic and dyspepsia). It can be applied to itchy skin. Oregano oil (*Origani vulgaris aetheroleum*) is used in aromatherapy and in dilute form for mouth hygiene and nasal congestion.

ORIGIN Europe to central Asia. Cultivated on a large scale for use as a culinary herb spice (mainly to flavour pizzas).

BOTANY Perennial herb (to 0.9 m); leaves hairy, opposite; flowers white or pink. *Origanum dictamnus* (dittany), *O. syriacum* (zatar) and *O. majorana* (marjoram) also have essential oil and similar uses.

CHEMISTRY Essential oil with **carvacrol** (40–70%), *p*-cymene and terpinene.

PHARMACOLOGY The oil has proven antimicrobial, spasmolytic and anti-inflammatory activity.

TOXICOLOGY The plant is non-toxic.

Origanum vulgare L. (Lamiaceae); *origan* (French); *Echter Dost* (German); *origano* (Italian); *orégano* (Spanish)

Orthosiphon aristatus
long-stamened orthosiphon

CLASSIFICATION TM: Europe, Asia. Comm.E+, ESCOP 1, PhEur8, HMPC.

USES & PROPERTIES Dried leaves (*Orthosiphonis folium*) are used to treat urological ailments in the form of a diuretic tea known as Java tea. It is used to treat inflammation of the bladder and kidneys. It may also be used with other herbs in urological preparations. Recommended daily dose: 6–12 g (taken as infusion, in doses of 2–3 g in 150 ml water).

ORIGIN Southeast Asia, Malaysia and Australia.

BOTANY Perennial herb (to 0.8 m); leaves opposite, hairy; flowers white/lilac; stamens protruding.

CHEMISTRY The herb is chemically complex, with methoxylated flavonoids (**sinensetin**, scutellarein), flavonol glycosides, organic acids (rosmarinic acid, tartaric acid) and diterpenes (orthosiphol A–C), potassium salts (3%) and essential oil (borneol, limonene, thymol and sesquiterpenoids).

PHARMACOLOGY Diuretic, antimicrobial, antioxidant and anti-inflammatory activity.

TOXICOLOGY No serious side effects are known.

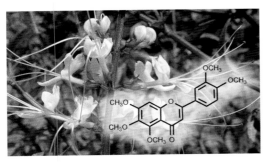

Orthosiphon aristatus (Blume) Miq. [= *O. stamineus* Benth.] (Lamiaceae); *moustache de chat* (French); *Katzenbart* (German); *kumis kutjing* (Indonesian); *tè de Giava* (Italian)

Paeonia lactiflora
white peony

CLASSIFICATION TM: Asia (China). Pharm., WHO 1. (*P. officinalis*: Pharm.).

USES & PROPERTIES Dried roots (*Paeonia radix*) are used in Chinese traditional medicine to treat digestive ailments (stomach cramps, liver problems), menstrual disorders (amenorrhoea, dysmenorrhoea) and also headache, dementia and vertigo. The daily dose is up to 15 g. Dried roots of the southern European *P. officinalis* (*Paeoniae radix rubra*) are no longer used to any extent but the petals (*Paeoniae flos*) are still included in herbal teas to provide colour.

ORIGIN Asia (China, Japan and India).

BOTANY Deciduous perennial herb (to 0.8 m); roots fleshy; leaves compound; flowers large, showy, usually white.

CHEMISTRY Roots contain a monoterpenoid glucoside called **paeoniflorin** (up to 5% of dry weight).

PHARMACOLOGY Paeoniflorin has analgesic, antipyretic, anti-inflammatory, sedative, uterus-contractant and vasodilatory activities.

TOXICOLOGY Non-toxic at recommended doses.

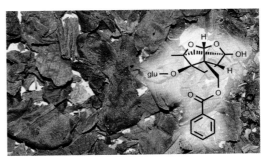

Paeonia lactiflora Pallas [= *P. albiflora*] (Paeoniaceae); *bai shao yao* (Chinese); *pivoine* (French); *Chinesische Pfingstrose*, *Päonie* (German); *peonia* (Italian)

Panax ginseng
ginseng • Asian ginseng

CLASSIFICATION TM: Asia (China), Europe. Comm.E+; WHO 1, 5; PhEur8, HMPC; clinical studies+. (*P. quinquefolius*: WHO 4).

USES & PROPERTIES Fresh or mostly dried roots (*Ginseng radix, ren shen*) have a long history of use in China as an adaptogenic tonic aimed at treating general weakness, fatigue, lack of stamina and declining concentration. *Panax quinquefolius* (American ginseng) is used in the same way.

ORIGIN East Asia (China). American ginseng: USA and Canada. Both cultivated on a large scale.

BOTANY Perennial herb; roots fleshy; leaves compound; flowers small, white; fruits fleshy, red.

CHEMISTRY Triterpenoid saponins, so-called ginsenosides, in a complex mixture: **ginsenoside Rg_1**, Rc, Rd, Rb_1, Rb_2 and Rb_0 are the main compounds. Also present are polyacetylenes.

PHARMACOLOGY Controlled clinical trials have proven that ginseng elevates and enhances mood, performance (physical and intellectual), immune response and convalenscence.

TOXICOLOGY No serious side effects are known.

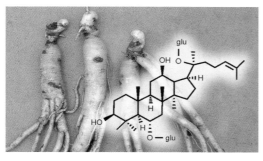

Panax ginseng C. A. Mey. (Araliaceae); *ginseng* (French); *Ginseng* (German); *ginseng* (Italian)

Papaver somniferum
opium poppy

CLASSIFICATION Neurotoxin, mind-altering, highly hazardous (Ib). TM: Asia, Europe. Pharm., PhEur8. MM: pure alkaloids.

USES & PROPERTIES Dried latex from unripe fruits (*Opium*) is traditionally used as intoxicant and nowadays as source of pure alkaloids for treating visceral spasms and intense pain, as well as cough, congestion, and the pain associated with colds and flu. Usual doses are 10–40 mg per day (morphine) and 30 mg every four hours (codeine).

ORIGIN Southwestern Asia.

BOTANY Annual (ca. 1.5 m); leaves grey; flowers large, terminal; fruit a multi-seeded capsule.

CHEMISTRY Raw opium contains isoquinoline alkaloids: **morphine** (10–12%), codeine (2.5–10%) and noscapine (= narcotine, 2–10%). Morphine is chemically converted to heroin (highly addictive).

PHARMACOLOGY Analgesic and euphoric (hallucinogenic). Morphine is converted to the less addictive codeine (antitussive, analgesic).

TOXICOLOGY Lethal dose: 50–75 mg heroin, 200–400 mg morphine (or 100–200 mg, parenteral).

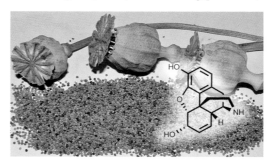

Papaver somniferum L. (Papaveraceae); *pavot somnifère* (French); *Schlafmohn* (German); *pavot officinal, papavero domestico* (Italian)

Phaseolus vulgaris
common bean • French bean

CLASSIFICATION Cell toxin (II–III). TM: Europe, South America. Pharm., Comm.E+, HMPC.
USES & PROPERTIES Dried bean pods, without seeds (*Phaseoli pericarpium*) are traditionally used as a weak diuretic (to treat urinary tract ailments, gout) and weak antidiabetic medicine. Included in herbal teas sold for kidney and bladder health. Uncooked seeds contain a toxic lectin that has to be destroyed by heating to make them edible.
ORIGIN Central and South America. An important vegetable and pulse crop, grown worldwide.
BOTANY Annual herb; stems twining (or short and branched in bushy cultivars); flower white or pink; fruit a narrowly oblong, multi-seeded pod.
CHEMISTRY Amino acids (e.g. asparagine, **arginine**, leucine, tyrosine), silicic acid and an alkaloid (trigonelline). The toxic lectin is phasin.
PHARMACOLOGY The diuretic activity is not yet explained (ascribed to silicic acid and arginine).
TOXICOLOGY Large quantities of the green pods or uncooked seeds may cause vomiting, diarrhoea and stomach pain.

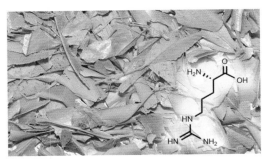

Phaseolus vulgaris L. (Fabaceae); *haricot* (French); *Gartenbohne* (German); *fagiolo* (Italian)

Phellodendron amurense
Amur cork tree

CLASSIFICATION TM: Asia (China). Pharm., WHO 4, clinical studies+ (berberine).
USES & PROPERTIES The inner bark (*huáng bò*) is one of the 50 major herbs in Chinese traditional medicine, used to treat meningitis, dysentery, pneumonia, liver cirrhosis and tuberculosis (also abdominal pain, diarrhoea, gastroenteritis, urinary tract inflammation and conjunctivitis).
ORIGIN East Asia. Invasive in parts of North America. The tree is planted in gardens and parks.
BOTANY Deciduous tree (to 12 m); bark corky; flowers small, yellow; fruit a fleshy black drupe.
CHEMISTRY Isoquinoline alkaloids (**berberine** is the main compound). Also sesquiterpene lactones and flavonols (e.g. amurensin).
PHARMACOLOGY Antimicrobial activity is usually ascribed to the DNA-intercalating berberine while anti-inflammatory effects may be due to phenolic compounds. Bark extracts are thought to be useful to prevent arthritis and some types of tumours.
TOXICOLOGY Use only under medical supervision. Do not exceed 3–10 g per day.

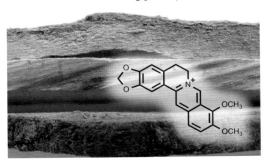

Phellodendron amurense Rupr. (Rutaceae) *huang bai* (Chinese); *Amur-Korkbaum* (German)

Phyllanthus emblica
Indian gooseberry • emblic

CLASSIFICATION TM: Asia (India, China).
USES & PROPERTIES The fresh or dried, sour-tasting fruit (without seed) is known as emblic or amlak and is used in India as a tonic to treat a wide range of ailments, especially fatigue, dyspepsia and diabetes. *Triphala* or "three fruits", the popular cure-all of Ayurvedic medicine, comprises the fruit pulp of amlak with that of *Terminalia bellirica* and *T. chebula*. In China it is called *yu gan zi* and used to treat throat infections.
ORIGIN Tropical Asia (often cultivated in parks).
BOTANY Deciduous tree (to 15 m); leaves very small, arranged in two rows; fruit a fleshy drupe.
CHEMISTRY Indolizidine alkaloids: **phyllantine** (= 4-methoxysecurinine). Lignans: phyllanthin and hypophyllanthin. Ellagitannins and gallotannins, polyphenols, flavonoids and triterpenoids.
PHARMACOLOGY Activities are mainly linked to the alkaloids (GABA receptor antagonists) and lignans. Amlak is described as rejuvenating, cooling and balancing.
TOXICOLOGY Fruits can be eaten raw or cooked.

Phyllanthus emblica L. (Phyllanthaceae); *yu gan zi* (Chinese); *groseille à maquereau indienne* (French); *Myrobalanenbaum* (German); *amla* (Hindi); *amalika* (Sanskrit); *mirobalano* (Spanish)

Physostigma venenosum
calabar bean • ordeal bean

CLASSIFICATION Neurotoxin, extremely hazardous (Ia). MM: alkaloids (physostigmine).
USES & PROPERTIES Physostigmine is used in modern therapy to treat glaucoma (it reduces intraocular pressure), Alzheimer's disease, delayed gastric emptying, short-term memory loss and orthostatic hypotension (a sudden drop in blood pressure when standing up or stretching). It is an antidote in atropine poisoning.
ORIGIN Tropical Africa (named after Calabar in Nigeria). A traditional ordeal poison (survival being proof of innocence, death proof of guilt).
BOTANY Woody climber (to 15 m); leaves trifoliate; flowers pink; fruit a large, 2–3-seeded pod.
CHEMISTRY Simple indole alkaloids: **physostigmine** and related compounds.
PHARMACOLOGY Physostigmine is a reversible cholinesterase inhibitor and acts as an indirect parasympathomimetic. It causes seizures, paralysis of the heart and death by asphyxiation.
TOXICOLOGY Extremely toxic. Physostigmine: LD_{50} (mouse) = 3 mg/kg (p.o.), 0.64 mg/kg (i.p.).

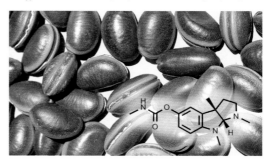

Physostigma venenosum Balf. (Fabaceae); *fève de Calabar* (French); *Calabarbohne* (German); *fava del Calabar* (Italian); *haba de Calabar, nuez esere* (Spanish)

Podophyllum peltatum
may apple • American mandrake

CLASSIFICATION Cell toxin (Ib); TM: North America. Pharm., Comm.E+. MM: lignans.

USES & PROPERTIES Dried rhizomes (*Podophylli peltati rhizoma*) have been used as purgative medicine (now obsolete) and also for removal of warts and condylomas. The herb and resin obtained from it (called podophyllin; *Podophylli resina*) are used in homoeopathy for liver and gall ailments. It is also a source of pure podophyllotoxin, used in modern medicine (only under strict supervision) to treat cancer and remove warts.

ORIGIN North America (*P. peltatum*) or Himalayan region (*P. hexandrum*, photo below).

BOTANY Perennial rhizomatous herb (to 0.5 m); leaves peltate, in one pair; flowers single, white.

CHEMISTRY Resin: up to 50% **podophyllotoxin**, a lignane used in cancer therapy. It is converted to more stable derivatives (e.g. etoposide).

PHARMACOLOGY The lignans inhibit tumours by inhibiting cell division.

TOXICOLOGY Podophyllotoxin is extremely poisonous: LD_{50} = 8.7 mg/kg (rat, i.v.).

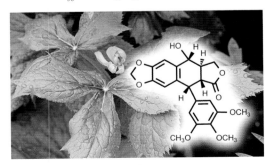

Podophyllum peltatum L. (Berberidaceae); *podophylle pelté, pomme de mai* (French); *Gewöhnlicher Maiapfel* (German); *podofillo* (Italian)

Pogostemon cablin
patchouli • patchouly

CLASSIFICATION TM: Asia (China, India and Malaysia).

USES & PROPERTIES Fresh or dried leaves (*Patchouli folium*) or the essential oil distilled from them (*Patchouli aetheroleum*) are used to treat colds and influenza, fever and headaches. It has the traditional reputation of being an aphrodisiac. The herb or the oil may be applied topically to soothe skin irritations and to repel insects and leeches. The oil is popular in aromatherapy (calming effect) and as an ingredient of cosmetics.

ORIGIN East Asia (India to Malaysia). The plant is commonly grown in gardens in all warm regions.

BOTANY Perennial aromatic herb (to 1 m); leaves soft, hairy; flowers pink (but plants rarely flower).

CHEMISTRY Essential oil with a unique sesquiterpene known as **patchoulol** (=patchouli alcohol). Also present in the oil are several monoterpenes, sesquiterpenes and phenylpropanoids.

PHARMACOLOGY Moisturising, antibiotic and protective effects on skin and mucosa.

TOXICOLOGY No side effects have been reported.

Pogostemon cablin (Blanco) Benth. [= *P. patchouli* Pellet.] (Lamiaceae); *patchouli* (French); *Patschulipflanze* (German); *patchouly* (Italian)

Polygala senega
senega • snakeroot senega

CLASSIFICATION TM: North America. Comm. E+, ESCOP 3, WHO 2, PhEur8. Homoeopathy.

USES & PROPERTIES The dried roots (senega root, *Polygalae radix*) are used in traditional medicine to treat coughs, chronic bronchitis, chronic asthma and emphysema. Infusions of 0.5–1 g of the dry root is taken three times a day. Decoctions are gargled for relief of throat infections.

ORIGIN North America (eastern and midwestern parts of the USA and most of Canada). Canadian Senega Indians used it to treat rattlesnake bites.

BOTANY Perennial herb (to 0.5 m); leaves sub-sessile; flowers with petaloid sepals, one of which has a small tuft (corona) at the tip.

CHEMISTRY Bidesmosidic saponins (up to 12%): senegasaponins A–D (the aglycone is **presenegin**). Also organic acids and methyl salicylate.

PHARMACOLOGY Expectorant, secretolytic, antitussive and anti-inflammatory activities are ascribed to the saponins and methyl salicylate.

TOXICOLOGY There are no serious side effects when used at recommended doses.

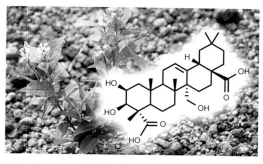

Polygala senega L. (Polygalaceae); *polygala sénéga* (French); *Senega Klapperschlangenwurzel* (German); *poligala, serpentella* (Italian)

Polygonum aviculare
knotweed • knotgrass

CLASSIFICATION TM: Europe, Asia (India, China). Pharm., Comm.E+, WHO 5, PhEur8.

USES & PROPERTIES Dried whole herb, including roots (*Polygoni avicularis herba*) is traditionally used for the treatment of upper respiratory tract infections (coughs, bronchial catarrh, sore throat). The recommended dose is 1.5 g of the herb, taken as infusion, three to five times a day. It has also been used as diuretic, topical haemostyptic and remedy for skin ailments.

ORIGIN Europe and Asia. It has become a cosmopolitan weed of cultivation.

BOTANY Wiry annual weed; stems slender; leaves small, with conspicuous sheathing stipules; flowers minute, white to pink.

CHEMISTRY The herb is rich in gallotannins and **condensed tannins** (= catechins) (3.6%). Also flavonols, coumarins, mucilage and salicic acid.

PHARMACOLOGY Expectorant, antibiotic and haemostyptic properties are ascribed to the tannins. Diuretic activity is linked to salicic acid.

TOXICOLOGY The herb is safe to use in low doses.

Polygonum aviculare L. (Polygonaceae); *renouée des oiseaux* (French); *Vogelknöterich* (German); *centinodia* (Italian)

Populus tremuloides
American aspen • Canadian aspen

CLASSIFICATION TM: North America (*P. tremuloides*); Europe. Comm.E+ (buds only, *P. tremula*).

USES & PROPERTIES Bark (*Populi cortex*), leaves (*Populi folium*) and buds (*Populi gemmae*) are used to treat the common cold, rheumatic conditions, cystitis and diarrhoea. *P. tremula* (photo below) is used in the same way.

ORIGIN North America. *P. tremula* and *P. nigra* (Europe and Asia) have similar uses. Poplar buds are obtained from balsam poplar (*P. balsamifera*) and balm of Gilead (*P. candicans*).

BOTANY Deciduous tree (to 20 m); leaves on long stalks, quivering; flowers inconspicuous.

CHEMISTRY Benzoyl esters of salicin: mainly **populin** (salicin-5-benzoate). Also salicin and salicortin (similar to willow bark), tannins and triterpenes. The buds are aromatic (essential oil).

PHARMACOLOGY When ingested, the salicin derivates are converted to salicylic acid (which has anti-inflammatory and analgesic activity).

TOXICOLOGY Non-toxic at low doses but general precautions apply.

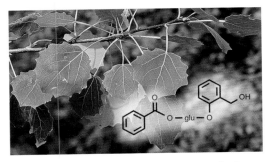

Populus tremuloides Michaux (Salicaceae); *peuplier faux-tremble* (French); *Amerikanische Espe* (German); *pioppo* (Italian)

Potentilla anserina
silverweed

CLASSIFICATION TM: Europe. Comm.E+.

USES & PROPERTIES Fresh or dried leaves and flowers (*Anserinae herba*) are used to treat non-specific diarrhoea and inflammation of the mouth and throat. Traditionally it has been used against colic, cramps, spasms and menstrual disorders. Topical uses include eczema, sores and bleeding haemorrhoids. A tea made from 2 g of the dry herb is taken three times a day. It can also be gargled for sore throat and mouth infections.

ORIGIN Europe, Asia and North America.

BOTANY Perennial ground-hugging herb; leaves pinnate, silver-hairy; flowers yellow.

CHEMISTRY Tannins (5–10%), based on **ellagic acid**. Flavonoids (flavonols, e.g. kaempferol, quercetin), proanthocyanidins, phenolic acids (caffeic acid, ferulic acid).

PHARMACOLOGY Astringent, anti-inflammatory and haemostyptic properties. These activities are due to tannins and other phenolics which also contribute to the spasmolytic properties.

TOXICOLOGY Safe to use at recommended doses.

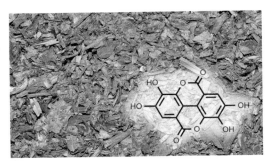

Potentilla anserina L. (Rosaceae); *herbe d'ansérine, argentine* (French); *Gänsefingerkraut* (German); *argentina anserina, potentilla* (Italian)

Potentilla erecta
tormentil

CLASSIFICATION   TM: Europe. Pharm., Comm. E+, PhEur8, HMPC.
USES & PROPERTIES   The dried rhizomes without roots (*Tormentillae rhizoma*) are a traditional remedy in Europe for diarrhoea, dysentery, gastroenteritis and enterocolitis. An infusion of 2–3 g is taken two or three times per day between meals. Infusions or tinctures may be gargled for sore throat and inflammations of the mouth. Externally it is applied to wounds, sores and inflamed skin.
ORIGIN   Europe (central and eastern regions).
BOTANY   Small perennial herb; leaves sessile, digitate, bright green, toothed, sparsely hairy.
CHEMISTRY   Tormentil rhizomes contain up to 20% catechin-type tannins. These include agrimoniin and other ellagitannins and catechin gallates. High levels of oligomeric proanthocyanidins (to 20%) and triterpene saponins (**tormentoside**).
PHARMACOLOGY   Tannins: antimicrobial, astringent and antidiarrhoeal activities. The triterpenes may be anti-inflammatory and antihypertensive.
TOXICOLOGY   Safe to use at prescribed doses.

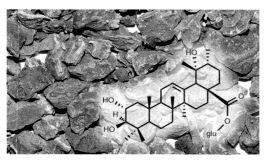

Potentilla erecta (L.) Räusch [= *P. tormentilla* Stokes] (Rosaceae); *tormentille* (French); *Blutwurz* (German); *tormentilla* (Italian)

Primula veris
cowslip

CLASSIFICATION   TM: Europe. Pharm., Comm. E+, ESCOP, PhEur8, HMPC.
USES & PROPERTIES   The rhizome and root (*Primulae radix*) are used as expectorants to treat coughs, bronchitis and catarrh of the nose and throat. Tea made from 0.2–0.5 g of the root is taken with honey, every two to three hours. The flowers with the calyx (*Primulae flos cum calycibus*) (2–4 g) are traditionally taken for relief of nervous conditions and headaches, and also as cardiac tonic.
ORIGIN   Europe and Asia.
BOTANY   Deciduous perennial herb; rhizome fleshy; leaves basal, wrinkled; flowers yellow.
CHEMISTRY   The flowers contain triterpene saponins and flavonoids. The roots have, in addition to triterpene saponins (aglycone: **primverogenin A** and B), also phenolic glycosides.
PHARMACOLOGY   Expectorant and secretolytic activities can be explained by the presence of saponins. The putative value in treating nervous conditions is not yet explained.
TOXICOLOGY   Do not exceed specified doses.

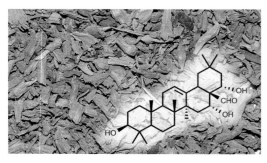

Primula veris L. [= *P. officinalis* (L.) Hill.] (Primulaceae); *primevère officinale, coucou* (French); *Wiesen-Schlüsselblume* (German); *primavera* (Italian)

Prunus africana
red stinkwood • pygeum

CLASSIFICATION TM: Africa, Europe. WHO 2, PhEur8, WHO 2, AHP, clinical studies+.

USES & PROPERTIES The bark (*Pygei africani cortex*) is used against benign prostate hyperplasia. Extracts of phytosterols are taken in doses of 100 mg per day for six to eight weeks. Commercial formulations often include saw palmetto fruits (*Serenoa repens*) and nettle roots (*Urtica dioica*).

ORIGIN Africa (tropical and subtropical parts). There are concerns about sustainable use.

BOTANY Evergreen tree (30 m), often with buttress roots; leaves glossy, dark green; flowers white, in spike-like racemes; fruit a small drupe.

CHEMISTRY Sitosterol and sitosterol glycosides (free and glycosylated **β-sitosterol** and campesterol). Also present are triterpenes and tannins.

PHARMACOLOGY Phytosterols appear to inhibit the binding of dihydrotestosterone in the prostate and may inhibit 5α-reductase and aromatase. Clinical evidence exists for efficacy in treating the symptoms of prostatitis.

TOXICOLOGY Phytosterols have a low toxicity.

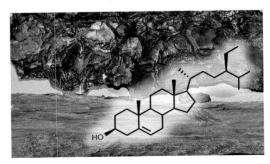

Prunus africana (Hook. f.) Kalkman [= *Pygeum africanum* Hook. f.] (Rosaceae); *pygeum* (French); *Pygeum africanum* (German); *pygeum* (Italian)

Prunus dulcis
almond

CLASSIFICATION Cell (respiratory) toxin, highly hazardous (Ib–II). TM: Europe. Pharm., PhEur8.

USES & PROPERTIES The ripe seeds (*Semen Amygdalae*) are the source of almond oil, which is used as dispersion agent for injections and as carrier oil in aromatherapy and cosmetics (also the source of "laetrile", a controversial cancer treatment).

ORIGIN Europe and Asia. Almonds are produced mainly in the USA (California), Spain and Italy.

BOTANY Deciduous tree (to 10 m); flowers white or pink; fruit velvety, single-seeded.

CHEMISTRY In common with other edible Rosaceae fruits, bitter almond seeds contain high levels of cyanogenic diglucosides (mainly **amygdalin**). Sweet almonds have only trace amounts.

PHARMACOLOGY Hydrogen cyanide (HCN) is released when amygdalin is cleaved by the enzyme emulsin. Small amounts are easily inactivated but large doses can be fatal due to respiratory arrest.

TOXICOLOGY The lethal dose of HCN in humans is 1 mg/kg (5–12 bitter almonds may be fatal for a child; 20–60 for an adult).

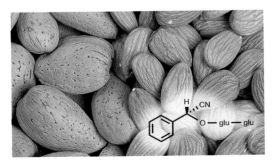

Prunus dulcis (Mill) D.A. Webb (Rosaceae); *amandier* (French); *Mandelbaum* (German); *mandorla* (Italian); *almendro* (Spanish)

Prunus laurocerasus
cherry laurel

CLASSIFICATION Cell toxin (Ib–II). TM: Europe.
USES & PROPERTIES The leaves were formerly used for treating mucosal infections of the mouth and throat. Cherry-laurel water is still occasionally used in Europe as a respiratory stimulant. It contains 0.1% hydrogen cyanide and was sometimes used for suicide and murder. The fruits resemble small cherries and are attractive to children.
ORIGIN Southeastern Europe and Asia Minor. It is grown in many parts of the world as a screen plant.
BOTANY Evergreen shrub (to 0.5 m); leaves large, glossy green; flowers white, in elongated clusters; fruit a small fleshy drupe, black when ripe.
CHEMISTRY The toxic compounds are cyanogenic glucosides: prunasin (up to 1.5% of fresh weight) in leaves; **amygdalin** (ca. 0.2%) in seeds.
PHARMACOLOGY The intact plant/leaf is harmless but damage (e.g. chewing) breaks the cells and expose the cyanogenic glucosides to enzymes, resulting in a release of hydrogen cyanide (HCN).
TOXICOLOGY HCN is extremely poisonous but fatal cases of poisoning by cherry laurel are rare.

Prunus laurocerasus L. (Rosaceae); *laurier-cerise* (French); *Kirschlorbeer, Lorbeerkirsche* (German); *lauroceraso* (Italian); *lauroceraso, laurel-cerezo* (Spanish)

Prunus spinosa
blackthorn • sloe

CLASSIFICATION Cell toxin (Ib–II). TM: Europe. Pharm., Comm.E+ (fruit only).
USES & PROPERTIES Ripe fruits, fresh or dried (*Pruni spinosae fructus*), are used for the treatment of mucosal infections of the mouth, gums and throat. The juice of fresh fruits or a tea made from 2–4 g of dried fruit are used as a gargle. Dried flowers (*Pruni spinosae flos*) have numerous traditional uses (as diaphoretic, diuretic, expectorant and mild laxative). They are used as an infusion of 1–2 g in a cup of boiling water, taken once or twice during the day or night.
ORIGIN Europe, western Asia and North America.
BOTANY Deciduous shrub (to 4 m); stems thorny; leaves dull green; flowers white; fruit fleshy, bluish black, resembling small plums.
CHEMISTRY Tannins, flavonoids and cyanogenic glucosides (**prunasin**) are the main compounds.
PHARMACOLOGY Tannins are astringent and often used for their antiseptic, antidiarrhoeal and anti-inflammatory activities.
TOXICOLOGY High doses can be dangerous.

Prunus spinosa L. (Rosaceae); *prunellier* (French); *Schlehdorn, Schlehe* (German); *prugnolo* (Italian)

Rhodiola rosea
roseroot • arctic root

CLASSIFICATION TM: Europe. Pharm., HMPC, clinical studies+.

USES & PROPERTIES The rhizomes have a long history of use by the Vikings and Russians: to increase strength, endurance and resistance to cold and disease; to promote fertility and a long life. It is today widely used (as standardised liquid extracts in 40% alcohol) as an adaptogenic tonic: to increase physical and mental endurance, and to moderate the symptoms of asthenia and old age. The fresh rhizome smells like attar-of-roses (hence *rosea*).

ORIGIN Arctic region (Scandinavia and Siberia).

BOTANY Perennial herb (to 0.6 m); leaves succulent; flowers yellow. (*Rhodiola* is sometimes considered to be a subgenus of *Sedum*.)

CHEMISTRY Phenylpropanoids (rosavins, 3% in liquid extracts): mainly **rosavin**, rosin and rosarin. Phenylethanols (0.8–1%): mainly salidroside (= rhodiolin). Also flavonoids, monoterpenes, acids.

PHARMACOLOGY The adaptogenic tonic uses listed above are supported by several clinical studies.

TOXICOLOGY No harmful side effects are known.

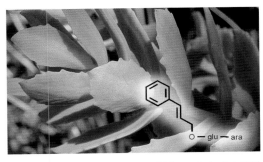

Rhodiola rosea L. [= *Sedum rosea* (L.) Scop.; *S. rhodiola* DC.] (Crassulaceae); *rhodiole rose* (French); *Rosenwurz* (German); *rodiola rosea* (Italian); *rosenrot* (Swedish)

Rhus toxicodendron
poison ivy • poison oak

CLASSIFICATION Cell toxin, strongly irritant, extremely hazardous (Ia).

USES & PROPERTIES Poison ivy is a common cause of allergic skin dermatitis (called poison ivy rash or *Rhus* dermatitis). Skin contact with the oleoresin that exudes from broken stems or leaves causes inflammation and blistering. About 350 000 people are affected in the USA each year (only a small percentage of the population is not allergic).

ORIGIN America (southern Canada to Guatemala) and eastern Asia (China and Japan).

BOTANY Variable: deciduous climber (vine), shrub or tree; leaves alternate, trifoliate, long-petiolate.

CHEMISTRY The resin is a mixture of several pentadecylcatechols (with C_{15} or C_{17} alkyl chains), collectively called urushiol (= toxicodendrin). An example is **urushiol I**.

PHARMACOLOGY Urushiol is absorbed within 10 minutes (wash immediately with soap and water).

TOXICOLOGY Urushiol is strongly irritant and extremely allergenic. It causes severe blistering and ulceration of the skin and mucosa.

Rhus toxicodendron L. [= *Toxicodendron pubescens* Mill.] (Anacardiaceae); *sumac grimpant* (French); *Giftsumach* (German); *edera velenosa* (Italian); *hiedra venenosa* (Spanish)

Ribes nigrum
blackcurrant

CLASSIFICATION TM: Europe. ESCOP 4, HMPC. DS: fresh fruits, juices, syrups and seed oil.
USES & PROPERTIES Dried leaves (*Ribis nigri folium*) are still popular as a diuretic tea (2–4 g of leaves, taken several times a day). Ripe fruits and juices are dietary supplements and functional food items during the cold and flu season. The seed oil is an alternative to evening primrose oil.
ORIGIN Central and eastern Europe. It is grown as a domestic and commercial crop in cold regions.
BOTANY Deciduous, aromatic shrub (ca. 1.5 m); leaves 5-lobed; flowers greenish red; fruit a multi-seeded berry, dark glossy purple, almost black.
CHEMISTRY Polyphenols and anthocyanins: the rutinosides and glucosides of delphinidin and cyanidin (**tulipanin** is a main compound). Flavonoids (0.5%), diterpenes and essential oil are also present. Leaves and especially fruits are rich in vitamin C. Seed oil contains 15% γ-linolenic acid.
PHARMACOLOGY Leaves have diuretic and hypotensive effects. The flavonoids are venotonics.
TOXICOLOGY Fruits and leaves are non-toxic.

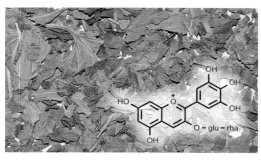

Ribes nigrum L. (Grossulariaceae); *cassis* (French); *Schwarze Johannisbeere* (German); *ribes nero* (Italian)

Ricinus communis
castor oil plant

CLASSIFICATION Cell toxin, extremely hazardous (Ia). TM: Europe, Africa. Pharm., PhEur8.
USES & PROPERTIES The seed oil (*Ricini oleum*), extracted by cold pressing (to avoid the water-soluble lectins) is a traditional purgative medicine. The seeds are used to oil the *mitad* (the pan used to prepare the traditional Ethiopian teff bread).
ORIGIN Africa (probably Ethiopia, the area of greatest genetic diversity). Widely naturalised.
BOTANY Robust shrub (to 4 m); leaves large; flowers terminal; fruit a 3-seeded capsule; seeds large.
CHEMISTRY The seeds contain ricin (a toxic lectin) and ricinine (a pyridine alkaloid). Ricin is one of the deadliest of all known natural substances. The laxative effect is due to **ricinoleic acid**.
PHARMACOLOGY Ricin inhibits ribosomal protein synthesis (injection causes fatal disruption of all vital organs). The alkaloid is also very toxic.
TOXICOLOGY The fatal dose of ricin in humans is 1 mg/kg (p.o.) or a mere 1 μg/kg (i.p.). Ingestion of five to six seeds can be lethal in children, 10 to 20 in adults (or 180 mg of milled seeds).

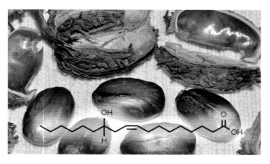

Ricinus communis L. (Euphorbiaceae); *ricin* (French); *Rizinus*, *Christuspalme* (German); *ricino* (Italian); *ricino* (Spanish)

Rosa canina
rose • dog rose

CLASSIFICATION TM: Europe. Pharm., Comm. E+ (petals only), ESCOP Suppl., PhEur8.

USES & PROPERTIES Dried ripe fruits, without the seeds (fruit pericarp; *Rosae pseudofructus*) are a traditional remedy for digestive ailments and are included in herbal teas for flavour. The seeds are traditionally used as diuretic and for treating urinary disorders. Dried rose petals from *R. centifolia* and *R. damascena* (*Rosae flos*) are used in mouth rinses.

ORIGIN Europe and Asia (naturalised in parts of North America and Africa). Fruits of *Rosa roxburghii* are used in China.

BOTANY Deciduous shrub (to 3 m); flowers single, pink; fruit fleshy, red (known as rose hips).

CHEMISTRY Rose hips are rich in **vitamin C** (to 2.4%). Also present are carotenoids (rubixanthin, lycopene and β-carotene), tannins, flavonoids, organic acids (and fatty acids such as GLA in seeds).

PHARMACOLOGY The fruits have demonstrated diuretic, anti-inflammatory and hypoglycaemic activity. Petals are antioxidant and astringent.

TOXICOLOGY No side effects at low doses.

Rosa canina L. (Rosaceae); *églantier* (French); *Hundsrose, Gemeine Heckenrose* (German); *rosa canina* (Italian)

Rosmarinus officinalis
rosemary

CLASSIFICATION TM: Europe. Pharm., Comm. E+, ESCOP, WHO 4, PhEur8, HMPC.

USES & PROPERTIES The dried leaves (*Rosmarini folium*) are used as a general tonic for digestive disturbances, headache and nervous complaints (2 g, 3 times/day). Essential oil (*Rosmarini aetherolium*) is used in ointments and bath oils for skin hygiene, improved blood circulation and minor pains. Internal use: 2 drops, up to 10 times/day).

ORIGIN Europe (Mediterranean region). Widely cultivated as a culinary herb and garden plant.

BOTANY Evergreen aromatic shrub (to 1 m); leaves narrow, silvery below; flowers usually blue.

CHEMISTRY Essential oil (2.5%): mainly 1,8-cineole, α-pinene and camphor. Of medicinal relevance is the presence of **rosmarinic acid**, bitter diterpenes, triterpene alcohols and flavonoids.

PHARMACOLOGY Rosemary herb is mildly spasmolytic, analgesic and choleretic. Rosmarinic acid has antioxidant, anti-inflammatory and antimicrobial activities. The flavonoids are venotonics.

TOXICOLOGY Non-toxic in small doses.

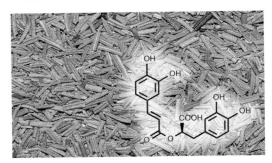

Rosmarinus officinalis L. (Lamiaceae); *romarin* (French); *Rosmarin* (German); *rosmarino* (Italian); *roméro* (Spanish)

Rubus fruticosus
bramble • blackberry

CLASSIFICATION   TM: Europe. Pharm., Comm. E+. (*R. idaeus*: HMPC).
USES & PROPERTIES   Enzymatically fermented and dried leaves (*Rubi fruticosi folium*) are used as astringent in European medicine to treat non-specific acute diarrhoea. It is also used against inflammation of the mouth and throat. Raspberry leaf (from *R. idaeus*) has similar uses (and to ease labour). Both are added to commercial herbal teas.
ORIGIN   Europe. A species complex and invasive weed in many parts of the world. Raspberry occurs naturally in Europe, Asia and North America.
BOTANY   Woody shrub (to 3 m); stems and leaves prickly; flowers white or pale pink; fruit black. Raspberry leaves are silver-hairy below and the red fruit separates from the stalk when ripe.
CHEMISTRY   Hydrolysable tannins (10%): **gallotannins, dimeric ellagitannins**. Also flavonoids.
PHARMACOLOGY   Astringent and antidiarrhoeal activities are typical of tannins. Raspberry leaf has uterotonic effects (contraction of uterine muscles).
TOXICOLOGY   No noteworthy side effects.

Rubus fruticosus L. (Rosaceae); *ronce noire* (French); *Brombeere* (German); *rovo* (Italian)

Ruscus aculeatus
butcher's broom

CLASSIFICATION   Cell toxin (III). TM: Europe. Pharm., Comm.E+, ESCOP, PhEur8, HMPC.
USES & PROPERTIES   Rhizomes and roots (*Rusci aculeati rhizoma*) are traditionally used to treat haemorrhoids and varicose veins. Extracts and pure compounds are included in ointments and suppositories (used for symptomatic relief). They are also ingredients of commercial mixtures for oral use, to treat venous and lymphatic vessel insufficiency.
ORIGIN   Europe and Asia.
BOTANY   Evergreen perennial shrublet (to 1 m); stems leaf-like (phylloclades); fruit a red berry.
CHEMISTRY   Steroidal saponins (to 6%): ruscin (monodesmosidic spirostane type) with its aglycone, **ruscogenin**; as well as ruscoside (bidesmodic furostane type) with its aglycone, neuruscogenin.
PHARMACOLOGY   The steroidal saponins have venotonic and anti-inflammatory activities.
TOXICOLOGY   *Ruscus* saponins are only slightly hazardous (no serious poisoning in humans) but may cause gastrointestinal disturbances.

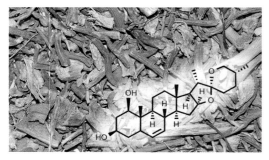

Ruscus aculeatus L. (Asparagaceae); *petit houx* (French); *Stechender Mäusedorn* (German); *pungitopo* (Italian)

Ruta graveolens
rue • herb of grace

CLASSIFICATION Cell toxin (Ib–II). TM: Europe. Pharm. (now obsolete).

USES & PROPERTIES The dried leaves (*Rutae folium*) or dried tops (*Rutae herba*) are traditional tonics, used for many ailments, including loss of appetite, dyspepsia, sore throat, fever, circulatory problems, venous insufficiency, high blood pressure, hysteria, menstrual disorders, abortion, arthritis, sprains, wounds and skin ailments.

ORIGIN Southern Europe. Widely cultivated as an ornamental plant. The Mediterranean Aleppo rue (*R. chalepensis*) is used in the same way.

BOTANY Aromatic shrub (to 1 m); leaves compound; flowers yellow; fruit a capsule; seeds black.

CHEMISTRY Coumarins, furanocoumarins, furanoquinoline alkaloids, flavonoids (**rutin**, 5%), essential oil (2-undecanone as major compound).

PHARMACOLOGY Antimicrobial, antispasmodic, anti-exudative, analgesic, ion-channel inhibiting. Rutin is a venotonic and capillary protectant.

TOXICOLOGY The furanocoumarins and alkaloids are mutagenic: avoid during pregnancy.

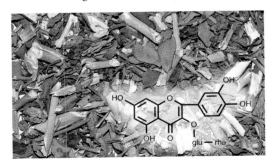

Ruta graveolens L. (Rutaceae); *rue* (French); *Weinraute* (German); *ruta* (Italian); *ruda común* (Spanish)

Salix alba
white willow

CLASSIFICATION TM: Europe. Pharm., Comm. E+, ESCOP 4, HMPC (*S. purpurea*) clinical studies+.

USES & PROPERTIES The dried bark from 2–3 year old branches (*Salix cortex*) is traditionally used for fever, influenza, headaches, rheumatism and minor pains (2–3 g, taken as tea, 3–4 × per day).

ORIGIN Europe and Asia. Basket willow (*S. purpurea*) and crack willow (*S. fragilis*) are also used.

BOTANY Deciduous tree (to 20 m); leaves silvery; flowers small, naked, in small spikes (catkins).

CHEMISTRY Phenolic glycosides (salicylates, salicortin, silicin); phenolic acids (chlorogenic acid) and oligomeric proanthocyanidins (1%).

PHARMACOLOGY Anti-inflammatory, analgesic, antipyretic and antirheumatic activies (supported by clinical studies). Salicortin is hydrolysed to salicin, which is converted by intestinal hydrolysis to saligenin. The latter is absorbed into the bloodstream and converted in the liver to **salicylic acid** (the active compound). Aspirin, the well-known analgesic, is the acetylated form of salicylic acid.

TOXICOLOGY Low toxicity; salicin is non-irritant.

Salix alba L. (Salicaceae); *saule blanc* (French); *Silberweide* (German); *salice bianco* (Italian)

Salvadora persica
mustard tree • toothbrush tree

CLASSIFICATION TM: Africa (dental care).
USES & PROPERTIES Pieces of root (ca. 10 mm in diameter and 100–200 mm long) are traditional toothbrush sticks. They are known as *miswak* (or *siwak*) and are commonly sold by vendors on local markets. The end of the stick is chewed until the fibres separate; the brush is rubbed across the teeth in vertical strokes. Root extracts ("peelu extracts") are ingredients of commercial toothpastes.
ORIGIN Africa (Namibia to North Africa) and Asia (Arabia to Pakistan and India).
BOTANY Shrub or small tree (to 5 m); leaves opposite, fleshy; flowers small; fruit a small red drupe.
CHEMISTRY Lignan glycosides, phenolic glycosides (syringin) and benzyl glucosinolates, the latter are enzymatically converted to volatile **benzyl isothiocyanate** (a mustard oil).
PHARMACOLOGY Strong antibacterial and antiplaque activities (supported by human studies). Benzyl isothiocyanate is active against gram-negative bacteria (those associated with periodontitis).
TOXICOLOGY Non-toxic in low doses.

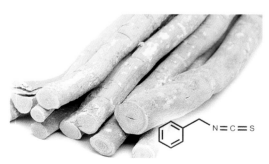

Salvadora persica L. (Salvadoraceae); *arak*, (Arabic); *salvadora persica* (French); *Senfbaum, Zahnbürstenbaum* (German); *salvadora persica* (Italian); *peelu* (Urdu)

Salvia divinorum
magic mint • diviner's sage

CLASSIFICATION Cell toxin, mind-altering (II). TM: Central America (Mexico).
USES & PROPERTIES The leaves have traditional medicinal uses in treating headache and digestive, rheumatic and urinary tract disorders. The plant is psychoactive and has been used by shamans for divining and healing rituals (it is perhaps the legendary *pipiltzintzintl* of the Aztecs). An illegal narcotic in some countries: leaves are chewed or smoked for an unusual psychoactive experience that lasts several minutes to one hour.
ORIGIN Central America (Oaxaca in Mexico). A sterile cultigen, sometimes grown in gardens.
BOTANY Perennial herb (to 2 m); leaves opposite; flowers 2-lipped, pale blue.
CHEMISTRY Diterpenoids (neoclerodane type): **salvinorin A** (the main active compound), with salvinorin B and other salvinorins (all inactive).
PHARMACOLOGY Salvinorin A is an agonist of the kappa opioid neuroreceptor (the only non-alkaloid to show this activity, at doses as low as 200 µg).
TOXICOLOGY Low toxicity; side effects unknown.

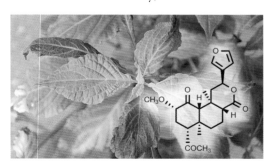

Salvia divinorum Epling & Játiva (Lamiaceae); *sauge des devins* (French); *Wahrsagesalbei, Zaubersalbei* (German); *salvia dei veggenti* (Italian)

Salvia officinalis
sage • garden sage

CLASSIFICATION TM: Europe. Comm.E+, ESCOP 2, WHO 5, PhEur8, HMPC.

USES & PROPERTIES The dried leaves (*Salviae folium*, *Salviae trilobae folium*) are traditionally used for a wide range of ailments, including digestive disorders, flatulence, diarrhoea, diabetes, excessive perspiration, night sweats, gingivitis and inflammation of the mucosa of the mouth and throat. The daily dose is 4–6 g of the herb or 0.3 g of the essential oil. Sage is a popular culinary herb.

ORIGIN Mediterranean Europe. Cultivated.

BOTANY Shrublet (to 0.6 m); leaves simple (basally lobed in trilobed sage, *S. fruticosa*); flowers blue.

CHEMISTRY Essential oil (3%): mainly thujone; diterpenoids: **carnosic acid** (= salvin), a preservative in food and oral hygiene products, and the structurally related carnosol (bitter value 14 000); triterpenes (oleanic acid) and rosmarinic acid.

PHARMACOLOGY Proven antispasmodic; antiseptic, carminative and antisudorific activities.

TOXICOLOGY Thujone is neurotoxic: avoid large amounts/chronic use (see *Artemisia absinthium*).

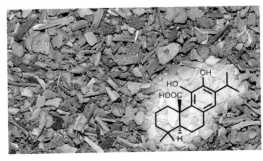

Salvia officinalis L. (Lamiaceae); *sauge officinale, sauge commune* (French); *Echter Salbei, Gartensalbei* (German); *salvia* (Italian); *salvia oficinal* (Spanish)

Sambucus nigra
elder • elderberry tree

CLASSIFICATION Cell toxin (III). TM: Europe, North America. Pharm., Comm.E+ (flowers), WHO 2 (flowers), PhEur8, HMPC.

USES & PROPERTIES The dried, sieved flowers, without stalks (*Sambuci flos*) or the fresh or dried fruits (*Sambuci fructus*) are traditionally used to treat colds, cough, catarrh and fever. An infusion of 3 g (flowers) or 10 g (fruits) is taken several times a day (or extracts and tinctures, such as elderberry wine). The flowers are an ingredient of herbal tea mixtures and herbal preparations while the fruits provide a natural colouring for food products.

ORIGIN Europe, Asia and North Africa.

BOTANY Shrub or small tree (to 6 m); leaves compound; flowers white; fruit a 3-seeded black drupe.

CHEMISTRY Flowers and fruits are rich in flavonoids: rutin (2%), quercitrin and hyperoside. Flowers: **chlorogenic acid** (3%) and triterpenoids; fruits: tannins (3%) and cyanidin glycosides.

PHARMACOLOGY Diaphoretic, anti-inflammatory (flowers, fruits); diuretic, mild laxative (fruits).

TOXICOLOGY Non-toxic when heated.

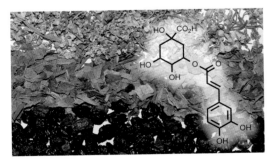

Sambucus nigra L. [Adoxaceae (formerly Caprifoliaceae)]; *grand sureau* (French); *Schwarzer Holunder* (German); *sambuco* (Italian); *sauco* (Spanish)

Sanguinaria canadensis
bloodroot

CLASSIFICATION Cell toxin (Ib–II). TM: North America. Pharm., clinical studies+.

USES & PROPERTIES The dried rhizome with the roots removed (*Sanguinariae canadensis rhizoma*) is traditionally used as expectorant, spasmolytic and emetic to treat asthma, bronchitis, cough, croup, laryngitis and pharyngitis. It is an ingredient of expectorants, cough syrups and stomachics and historically also oral hygiene products such as mouth rinses and toothpastes.

ORIGIN North America. It is an ornamental plant.

BOTANY Perennial herb (to 0.4 m); rhizomes and stems with red sap; leaves rounded, lobed; flowers large, white; fruit a bilocular capsule.

CHEMISTRY Isoquinoline alkaloids (9%): mainly **sanguinarine** (50%) and several others.

PHARMACOLOGY Sanguinarine is strongly antibiotic and anti-inflammatory. It counteracts dental plaque and gingivitis (showed in clinical studies).

TOXICOLOGY Sanguinarine: LD_{50} (mouse) = 19.4 mg/kg (i.v.), 102 mg/kg (s.c.). It is no longer used to any extent in oral rinses because of side effects.

Sanguinaria canadensis L. (Papaveraceae); *sanguinaire du Canada* (French); *Kanadische Blutwurzel* (German); *sanguinaria* (Italian)

Sanguisorba officinalis
burnet • greater burnet • garden burnet

CLASSIFICATION TM: Europe, Asia. PhEur8.

USES & PROPERTIES The dried rhizome and root (*Sanguisorbae rhizoma et radix*) and sometimes the dried aboveground parts (*Sanguisorbae herba*) are traditionally used to stop bleeding (hence the generic name). It is an effective remedy for acute diarrhoea and has been used to treat inflammation of the mouth and throat, ulcerative colitis, uterine bleeding, burns, wounds, ulcers and eczema.

ORIGIN Europe and Asia. Salad burnet (*S. minor*) is a culinary herb with similar medicinal uses. Both species are commonly grown in herb gardens.

BOTANY Erect perennial herb (to 1 m); leaves pinnate; flowers small, in dense, oblong heads.

CHEMISTRY Ellagitannins: sanguiin H-6 is a major compound (an isomer of agrimoniin with **sanguisorbic acid** ester groups as linking units); also saponins (sanguisorbin) and proanthocyanidins.

PHARMACOLOGY The astringent, haemostyptic, antihaemorrhoidal, antimicrobial and anti-inflammatory activities are ascribed to the tannins.

TOXICOLOGY No serious side effects are known.

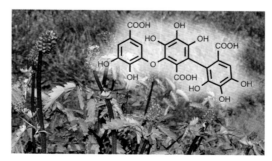

Sanguisorba officinalis L. (Rosaceae); *sanguisorbe*; *grande pimprenelle* (French); *Großer Wiesenknopf* (German); *sorbastrella* (Italian)

Syzygium aromaticum
clove tree

CLASSIFICATION TM: Europe, Asia (India). Comm.E+., WHO 2, PhEur8, HMPC.

USES & PROPERTIES Dried flower buds, known as cloves (*Caryophylli flos*) or the essential oil distilled from them (clove oil; *Caryophylli aetheroleum*) are traditionally used against toothache (a few drops of pure oil) and to treat inflammation of the mouth and throat, nausea and gastrointestinal disorders (diluted to 5%). Externally applied against rheumatism and myalgia. A culinary spice.

ORIGIN Southeast Asia (Moluccas Islands).

BOTANY Evergreen tree (to 12 m); leaves glossy; flowers white, with prominent stamens.

CHEMISTRY Essential oil in high yields (15–20%), with **eugenol** (85%), eugenyl acetate (to 15%) and β-caryophyllene (to 7%). Also flavonoids, tannins, phenolic acids and triterpenes.

PHARMACOLOGY The oil has local anaesthetic, antiseptic, antispasmodic and carminative activities. Eugenol is analgesic and anti-inflammatory.

TOXICOLOGY Cloves are not toxic in small amounts but the oil can cause allergic reactions.

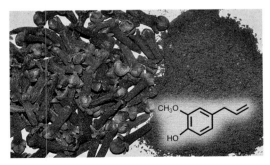

Syzygium aromaticum (L.) Merr.& Perry (Myrtaceae); *giroflier* (French); *Gewürznelkenbaum* (German); *chiodi di garofano* (Italian); *clavero* (Spanish)

Tagetes minuta
Mexican marigold • wild marigold

CLASSIFICATION TM: South America.

USES & PROPERTIES Dried aboveground parts are traditionally used as medicinal tea for the common cold, infections of the upper and lower respiratory tract, digestive ailments, stomach upsets, diarrhoea and liver ailments. Used since ancient times as culinary herb and flavourant in beverages and condiments (known as black mint). The oil is used in perfumery and as flavour compound in many major food products, including cola drinks.

ORIGIN South America (southern parts). It has become a weed in most parts of the world.

BOTANY Erect annual herb (1–2 m); leaves pinnate, glandular; flower heads small, yellow.

CHEMISTRY Essential oil with **dihydrotagetone,** β-ocimene and tagetone as main compounds. Also present are several thiophenes.

PHARMACOLOGY Anti-inflammatory, bronchodilatory, hypotensive, spasmolytic and tranquilising effects (shown in animal studies). The thiophenes are also antiviral and antibacterial.

TOXICOLOGY No ill effects; use in moderation.

Tagetes minuta L. (Asteraceae), *tagète, tagette* (French); *Mexikanische Studentenblume* (German); *cravo de defuncto* (Portuguese); *anisillo, huacatay* (Spanish)

Tamarindus indica
tamarind • Indian date

CLASSIFICATION TM: Africa, Asia (India). Pharm. DS: functional food.

USES & PROPERTIES The fleshy, reddish brown fruit mesocarp (*Tamarindorum pulpa*) is traditionally eaten as a mild laxative. It may be used as a general tonic to improve appetite and digestion. Extracts and infusions are also used in India as general tonics to treat fever, liver conditions and bile ailments. It is widely used in the food industry (e.g. in drinks, chutneys and condiments).

ORIGIN Northeast Africa (Ethiopia and Sudan). It spread from here to India and the rest of Asia. Tamarind means "Indian date" (*tamar* = date).

BOTANY Evergreen tree (to 25 m); leaves pinnately compound; flowers yellow; fruit a fleshy pod.

CHEMISTRY The edible but sour fruit pulp contains pectins, sugars and organic acids (12–15%): **tartaric acid**, malic acid and citric acid. Also minor aromatic substances (cinnamates).

PHARMACOLOGY The pulp has a mild laxative effect and is also astringent and mildly antiseptic.

TOXICOLOGY The fruit pulp is non-toxic (edible).

Tamarindus indica L. (Fabaceae); *tamarinier* (French); *Tamarinde* (German); *tamarindo* (Italian)

Tanacetum parthenium
feverfew

CLASSIFICATION TM: Europe. Pharm., ESCOP 2, WHO 2, PhEur8, HMPC, clinical studies+.

USES & PROPERTIES Aboveground parts (*Tanaceti parthenii herba*) are mainly used as a migraine prophylaxis but traditionally also to treat fever, rheumatism, skin ailments and gynaecological disorders. In a clinical study, a daily dose of 100 mg dry herb per day (equivalent to 0.5 mg parthenolide) reduced the number of migraine attacks.

ORIGIN Southern Europe and western Asia. It is commonly grown as a garden plant.

BOTANY Perennial herb (to 0.5 m); leaves lobed, aromatic; flower heads yellow and white.

CHEMISTRY Sesquiterpene lactones: **parthenolide** is the main compound. Also an essential oil (with camphor, chrysanthenyl acetate, camphene).

PHARMACOLOGY The lactones inhibit the formation of prostaglandins and leucotrienes; anti-inflammatory, spasmolytic and antimicrobial.

TOXICOLOGY Feverfew is not toxic but may cause side effects in sensitive persons: skin rashes, mouth ulcers, indigestion and stomach pain.

Tanacetum parthenium (L.) Sch. Bip. (Asteraceae); *grande camomille* (French); *Mutterkraut* (German); *partenio* (Italian); *matricaria* (Spanish)

Taraxacum officinale
dandelion

CLASSIFICATION TM: Europe. Pharm., Comm. E+, ESCOP 2, HMPC.

USES & PROPERTIES The whole herb (roots and leaves) collected just before flowering (*Taraxaci radix cum herba*) used as a diuretic, mild choleretic, bitter tonic and as supportive treatment in case of liver and gall bladder ailments. In traditional medicine it is a mild laxative and remedy for arthritis and rheumatism; externally for skin ailments. Infusions of 4–10 g are taken 3×/day.

ORIGIN Europe, Asia and North America. A diverse species complex and cosmopolitan weed.

BOTANY Perennial herb with a fleshy taproot; all parts with milky latex; leaves toothed; flower heads solitary; fruit a small, wind-dispersed achene.

CHEMISTRY Bitter sesquiterpene lactones: tetrahydroridentin B and **taraxacolide β-*D*-glucoside**; triterpenes (taraxasterol); sterols; flavonoids.

PHARMACOLOGY Bitter lactones are probably responsible for diuretic and cholagogic activities.

TOXICOLOGY No toxic or adverse effects. The latex may cause dermatitis after repeated exposures.

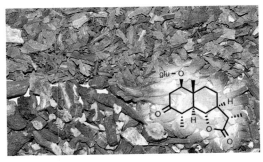

Taraxacum officinale Weber ex Wigg. (Asteraceae); *pissenlit, dent de lion* (French); *Gemeiner Löwenzahn* (German); *taraxaco* (Italian); *diente de leon* (Spanish)

Taxus baccata
yew • English yew

CLASSIFICATION Cell toxin, extremely hazardous (Ia). MM: paclitaxel, clinical studies+.

USES & PROPERTIES The bark (originally) and nowadays the leaves of *Taxus* species are extracted for starting materials to produce paclitaxel analogues. Paclitaxel was originally isolated from the Pacific yew (*T. brevifolia*). Ca. 110–250 mg/m² per body per day are used to treat ovarian and breast cancer (at 3–4 week intervals).

ORIGIN Europe (Mediterranean). (*T. brevifolia* occurs on the west coast of North America.)

BOTANY Evergreen tree (to 20 m); leaves linear; cone small, surrounded by a red, non-toxic aril.

CHEMISTRY Diterpene pseudoalkaloids such as **paclitaxel** (Taxol®) and structural analogues (e.g. docetaxel or Taxotere®) are produced by semisynthesis from the diterpenes extracted from leaves.

PHARMACOLOGY Taxol® and related compounds are spindle poisons that stop cell division by preventing the depolymerisation of microtubules.

TOXICOLOGY Extremely poisonous: 50 g of leaves can be fatal. Not suitable for self-medication.

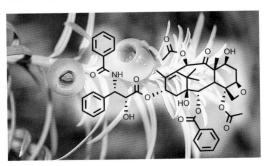

Taxus baccata L. (Taxaceae); *if* (French); *Eibe* (German); *tasso* (Italian); *tejo* (Spanish)

Terminalia chebula
black myrobalan • chebulic myrobalan

CLASSIFICATION TM: Asia (India). WHO 4.
USES & PROPERTIES The dried, ripe fruit rind, known as black chebulic or black myrobalan (*Myrobalani fructus*; *abhaya* in Sanskrit) is used as a panacea to treat ongoing colds, cough, flatulence, constipation, diarrhoea, piles, lack of libido and memory loss. It is used externally for wound healing and as gargle for inflammation in the mouth and throat. In Ayurvedic medicine it is one of the three components of *triphala* (a general tonic).
ORIGIN Asia (India to China).
BOTANY Deciduous tree (to 25 m); leaves simple; flowers yellow; fruit a large orange-brown drupe.
CHEMISTRY The main compounds are tannins, coumarins (e.g. chebulin) and triterpenes. **Chebulic acid** was isolated from the ripe fruits. Together with ellagitannins it occurs as chebulagic acid and chebulinic acid.
PHARMACOLOGY Digestive, anti-inflammatory, anthelmintic, cardiotonic, aphrodisiac and restorative properties have been ascribed to the fruits.
TOXICOLOGY No serious side effects are known.

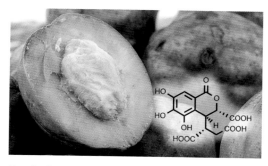

Terminalia chebula Retz. (Combretaceae); *myrobalan noire* (French); *Chebula-Myrobalane* (German); *haritaki* (Sanskrit)

Theobroma cacao
cacao

CLASSIFICATION TM: Europe, C and S America.
USES & PROPERTIES Ripe, fermented seeds (*Cacao semen*) were used to prepare ritual drinks with a reputation of curing stomach ailments, bronchitis and catarrh, but also acting as stimulants and aphrodisiacs. Cacao butter (*Cacao oleum*) was once a carrier for suppositories and is used in ointments and cosmetics. Cacao solids are used for chocolate, chocolate drinks and dietary supplements.
ORIGIN Tropical Central and South America. Cultivated also in Africa and Asia.
BOTANY Evergreen tree (4–8 m); leaves alternate; fruit large, multi-seeded; seeds 25 mm long.
CHEMISTRY The seeds contain triglycerides (50%), flavonoids, tannins, and alkaloids [mainly **theobromine** (3%) and caffeine (0.3%)].
PHARMACOLOGY Theobromine is stimulant, diuretic, antitussive and reduces high blood pressure. Flavanols have antioxidant and vasotonic effects.
TOXICOLOGY Habitual use of large doses lead to appetite loss, nausea, anxiety and withdrawal headaches. Chocolate is toxic to dogs and cats.

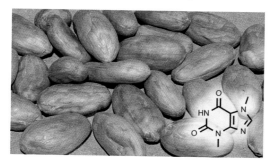

Theobroma cacao L. [Malvaceae formerly Sterculiaceae)]; *cacaotier, cacaoyer* (French); *Kakaobaum* (German); *cacao* (Italian); *cacao real* (Spanish)

Thevetia peruviana
yellow oleander

CLASSIFICATION Cell toxin (Ib). TM: Europe, North America (now obsolete).

USES & PROPERTIES The seeds (*Thevetiae semen*) are very poisonous and have resulted in accidental deaths (also suicide and murder). Extracts and pure compounds have been used to treat the symptoms of heart insufficiency. A mixture of heart glycosides called *Thevetin* was once sold as a heart tonic. Powdered seeds were used as rat poison.

ORIGIN Central and South America. Yellow oleander is a popular garden shrub in all warm regions.

BOTANY Shrub (to 10 m); leaves glossy, with white latex; flowers yellow; fruit a 2–4-seeded capsule.

CHEMISTRY A large number of heart glycosides (cardenolides; 5% in seeds): **thevetin A** and B, and peruvoside are the main compounds.

PHARMACOLOGY The cardenolides inhibit Na^+, K^+-ATPase, resulting in a disruption of signal transduction to the heart muscle cells.

TOXICOLOGY All parts of the plant are toxic but the seeds are extremely poisonous: four can be fatal in children, eight to 10 in adults.

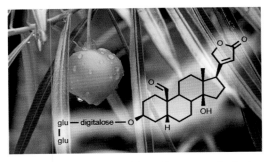

Thevetia peruviana (Pers.) K. Schum. (Apocynaceae); *arbre à lait, laurier jaune* (French); *Gelber Oleander, Gelber Schellenbaum* (German); *oleandro giallo* (Italian)

Thuja occidentalis
American arbor-vitae • white cedar

CLASSIFICATION Cell toxin, neurotoxin, mind-altering (Ib). TM: North America.

USES & PROPERTIES The flattened branches with scale leaves (*Thujae summitates*) or the essential oil (*Thujae aetheroleum*) are traditionally used to treat fever, colds, bronchitis, cystitis, rheumatism and headaches (formerly to induce abortion). Extracts and ointments are used to treat warts, skin rashes, rheumatic pain and neuralgia.

ORIGIN North America. A popular garden tree.

BOTANY Evergreen tree (to 20 m); branches flattened; leaves minute, scale-like; cones small. *T. orientalis* (photo below) is chemically similar.

CHEMISTRY Essential oil with up to 65% thujone (both **α-thujone** and β-thujone); also polysaccharides, sesquiterpenes and flavonoids.

PHARMACOLOGY A traditional decongestant and diuretic, thought to have immune-stimulant activity. Thujone is a cumulative neurotoxin.

TOXICOLOGY Chronic use of thujone may lead to hallucinations, depression and epileptic seizures. α-Thujone: $LD_{50} = 87.5$ mg/kg (mouse, s.c.).

Thuja occidentalis L. (Cupressaceae); *thuya d'occident, thuya américain* (French); *Abendländischer Lebensbaum* (German); *thuja* (Italian); *tuya* (Spanish)

Thymus vulgaris
thyme • garden thyme

CLASSIFICATION TM: Europe, Africa. Comm. E+, ESCOP 1, WHO 1, 5, PhEur8, HMPC.

USES & PROPERTIES Dried leaves and flowers (*Thymi herba*) and volatile oil (*Thymi aetheroleum*) are used against respiratory ailments (coughs, colds, bronchitis) and digestive disturbances (upset stomach, stomach cramps, lack of appetite). Tea made from 1–4 g is taken several times a day. Extracts are gargled for sore throat or mucosal infections of the mouth and can be applied to the skin to treat minor wounds and rashes. The oil or extracts are used in commercial products.

ORIGIN Europe. Spanish thyme (*T. zygis*) is an acceptable alternative; also other species (grown in herb gardens) are used in traditional medicine.

BOTANY Perennial shrublet (to 0.3 m); leaves opposite, greyish green; flowers pale to dark violet.

CHEMISTRY Essential oil with **thymol** (to 50%) and carvacrol; rosmarinic acid; flavonoids.

PHARMACOLOGY Oil (and thymol) is antibiotic; also expectorant and spasmolytic.

TOXICOLOGY Safe in small doses (culinary herb).

Thymus vulgaris L. (Lamiaceae); *thym* (French); *Echter Thymian* (German); *timo* (Italian); *tomillo* (Spanish)

Tilia cordata
lime • linden

CLASSIFICATION TM: Europe. Pharm., Comm. E+, WHO 5, PhEur8, HMPC.

USES & PROPERTIES Dried flowers with the bracts (*Tiliae flos*) are traditionally used as herbal tea, mainly to treat coughs, colds and fever but also influenza, sore throat, bronchitis, digestive ailments, nausea, hysteria and palpitations. Externally it is used against itchy skin problems. The daily dose is 2–4 g. The herb or extracts (also from the sapwood) are included in commercial products.

ORIGIN Europe. Large-leaved lime (*T. platyphyllos*) is an alternative source. Both are garden trees.

BOTANY Deciduous tree (to 30 m); leaves heart-shaped; flowers subtended by a large bract.

CHEMISTRY Rich in mucilage (arabino-galactans, 10%), proanthocyanidins, gallocathechins, phenolic acids and flavonoids (e.g. quercetin glycosides and **tiliroside**, a coumaric acid ester).

PHARMACOLOGY Diaphoretic activity; also mild sedative, antispasmodic, emollient, antioxidant, diuretic and astringent.

TOXICOLOGY No side effects are known.

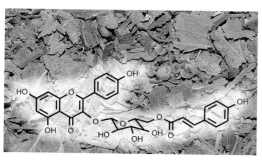

Tilia cordata Mill. [Malvaceae (formerly Tiliaceae)]; *tilleul à petites feuilles* (French); *Winterlinde* (German); *tiglio* (Italian)

Trifolium pratense
red clover

CLASSIFICATION TM: Europe, Asia. DS: phytoestrogens.

USES & PROPERTIES Flowering aboveground parts (*Trifolii pratensis flos*) are used in Ayurvedic medicine as expectorant, antispasmodic and anti-inflammatory medicine. It is used to extract isoflavones, marketed as dietary supplements and as an alternative to hormone replacement therapy. It is used to treat menopausal symptoms and is claimed to prevent cancer and osteoporosis.

ORIGIN Europe and Asia. A cultivated pasture and common weed in many parts of the world.

BOTANY Short-lived perennial herb (to 0.8 m); leaves trifoliate, with pale marks; flowers pink.

CHEMISTRY Isoflavones: mainly **formononetin** and biochanin A, with some genistein and daidzein. Also cyanogenic glucosides and salicylic acid.

PHARMACOLOGY Isoflavonoids can mimic the effects of endogenous oestrogens. Medical claims are not unambiguously supported by clinical data.

TOXICOLOGY Non-toxic when used in small doses but side effects are possible in cancer patients.

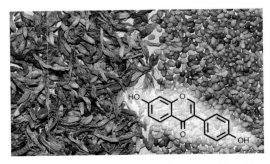

Trifolium pratense L. (Fabaceae); *trèfle commun* (French); *Rotklee, Wiesenklee* (German); *trifoglio* (Italian)

Trigonella foenum-graecum
fenugreek

CLASSIFICATION TM: Africa, Europe, Asia. Pharm., Comm.E+, WHO 3, PhEur8, HMPC, clinical studies+. DS: functional food.

USES & PROPERTIES Ripe, dried seeds (*Foenugraeci semen*) are used as a digestive tonic for loss of appetite, anorexia and weight loss. Daily dose: 3–18 g. Functional food, used in supportive treatment of high cholesterol and diabetes. Seeds are applied as poultice, oils or ointments for relief of boils, ulcers, eczema and inflamed skin. The seeds and leaves have many uses as food and spice.

ORIGIN Mediterranean region, North Africa and western Asia. Cultivated as a crop since antiquity.

BOTANY Annual herb (to 0.5 m); leaves trifoliate; flowers white; fruit a many-seeded, pointed pod.

CHEMISTRY Mucilage (to 45%, galactomannans); proteins (to 30%); steroidal saponins (glycosides of **diosgenin** and yamogenin); alkaloids; peptides.

PHARMACOLOGY Hypoglycaemic and cholesterol-lowering effects supported by clinical studies (also expectorant, anti-inflammatory, uterotonic).

TOXICOLOGY Non-toxic (edible).

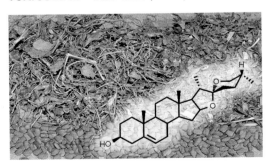

Trigonella foenum-graecum L. (Fabaceae); *fénugrec* (French); *Griechischer Bockshornklee* (German); *fieno-greco* (Italian): *fenugreco* (Spanish)

Tropaeolum majus
nasturtium

CLASSIFICATION TM: Central and South America, Europe. Comm.E+. DS: functional food.
USES & PROPERTIES Fresh or dried leaves (*Tropaeoli herba*) are traditionally used as a natural antibiotic to help clear up infections of the respiratory and urinary tracts. Leaves can also be applied in the form of a poultice as a counter-irritant in case of rheumatic or muscular pain and to treat candida and other fungal infections.
ORIGIN South America (Peru). Nasturtium is a popular garden plant, grown for its edible flowers.
BOTANY Spreading perennial (to 1.5 m); stems fleshy; leaves peltate; flowers large, with a distinctive nectar spur; seeds round, fleshy.
CHEMISTRY The main active compound is benzyl isothiocyanate (= benzyl mustard oil). It is formed through enzymatic hydrolysis from **glucotropaeolin** (= benzyl glucosinolate) in the intact plant.
PHARMACOLOGY Mustard oils bind to proteins, hence their antiviral, antifungal and antibacterial activities. Benzyl mustard oil is a severe irritant.
TOXICOLOGY All parts are edible (used in salads).

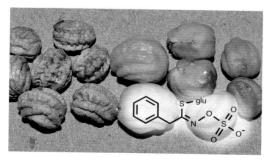

Tropaeolum majus L. (Tropaeolaceae); *capucine grande, cresson d'Inde* (French); *Große Kapuzinerkresse* (German); *nasturzio* (Italian)

Turnera diffusa
damiana

CLASSIFICATION Cell toxin, stimulant (III). TM: Central America. Pharm.
USES & PROPERTIES Dried leaves (*Damianae folium*) are used since the time of the Mayas as stimulant, general tonic and aphrodisiac. An infusion of 2–4 g of dry leaves is taken 3 × per day to counteract stress, depression and fatigue, with putative benefits in impotence, loss of libido, prostate problems and menstrual disturbances. Leaves can be smoked with effects similar to marijuana. Extracts are used in urinary products and tonics.
ORIGIN Tropical America (Mexico, southern California and the Caribbean Islands).
BOTANY Aromatic shrub (to 2 m); leaves small, prominently veined; flowers yellow.
CHEMISTRY Cyanogenic glucoside (**tetraphyllin B**); arbutin (0.7%); essential oil with α-pinene, β-pinene, calamene, α-copaene and others.
PHARMACOLOGY The aphrodisiac use is claimed to be supported by animal studies but the tonic and euphoric activities are not yet explained.
TOXICOLOGY Poorly known (but used as tea).

Turnera diffusa Willd. [= *Damiana diffusa* var. *aphrodisiaca* (L.F. Ward.) Urb.] [Passifloraceae (formerly Turneraceae)]; *damiana* (French); *Damiana* (German)

Vaccinium myrtillus
bilberry

CLASSIFICATION TM: Europe, North America. Comm. E+, WHO 4, PhEur8. DS: dry fruits.
USES & PROPERTIES Dried, ripe berries (*Myrtilli fructus*) are mainly used to treat diarrhoea in children, as well as inflammation of the mouth and throat. Daily dose: 20–60 g of dried berries. The fruits and leaves have many traditional uses.
ORIGIN Europe, Asia and North America. The American blueberry (highbush blueberry) is *V. corymbosum* (much lower levels of anthocyanins).
BOTANY Shrublet (to 0.8 m); stems angular; flowers small, white; fruit a bluish black berry, 6 mm in diameter (to 12 mm in *V. corymbosum*).
CHEMISTRY Anthocyanins (7.5%), mainly **myrtillin** (= delphinidin 3-*O*-glucoside). *V. corymbosa*: malvidin/petunidin. Also condensed tannins (12%); proanthocyanidins; flavonoids; iridoid glucosides.
PHARMACOLOGY Antidiarrhoeal activity, ascribed to the tannins. Lipid-lowering and hypoglycaemic effects have been demonstrated (also antimicrobial, venotonic, wound-healing, antioxidant).
TOXICOLOGY Edible (even in large amounts).

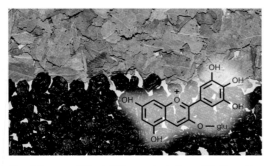

Vaccinium myrtillus L. (Ericaceae); *airelle myrtille* (French); *Heidelbeere, Blaubeere* (Germany); *mirtillo* (Italian); *arandano commun* (Spanish)

Vaccinium vitis-idaea
cowberry • lingonberry

CLASSIFICATION TM: Europe, Asia, North America. DS: fruits and fruit juice.
USES & PROPERTIES Ripe fruits, fruit extracts or fruit juice are used as dietary supplements to prevent oxidative stress and reduce the risk of cardiovascular diseases. Leaves are traditionally used as diuretic and urinary tract antiseptic.
ORIGIN Europe, Asia and North America.
BOTANY Mat-forming evergreen shrublet; leaves glossy; flowers white, waxy; fruit a red berry, 6–8 mm in diameter.
CHEMISTRY Fruits are rich in anthocyanins (mainly **cyanidin 3-*O*-galactoside**). Also vitamin C (22 mg per 100 g), minerals and organic acids. Leaves have high levels of tannins and anthocyanins.
PHARMACOLOGY A human study showed that the cyanidin galactoside (partly metabolised) is excreted in urine 4–8 hours after ingestion. Anthocyanins have antioxidant and radical-scavenging activities. Indications of anti-inflammatory, anti-obesity and antidiabetic effects.
TOXICOLOGY The fruits are non-toxic (edible).

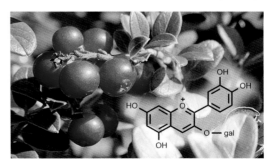

Vaccinium vitis-idaea L. (Ericaceae); *airelle rouge, myrtille rouge* (French); *Kronsbeeren, Preiselbeere* (German); *mirtillo rosso* (Italian)

Valeriana officinalis
valerian • common valerian

CLASSIFICATION Cell toxin, mind-altering (III). TM: Europe. Comm.E+, ESCOP 4, WHO 1; PhEur8, HMPC, clinical studies+.
USES & PROPERTIES Rhizomes and roots (*Valerianae radix*) are important commercial sedatives and tranquillisers, used in many products to treat restlessness, anxiety, sleeplessness and the symptoms of menopause and premenstrual syndrome. Daily dose: 10 g (2–3 g, up to 5 × per day, as tea).
ORIGIN Europe, Asia (naturalised in N. America).
BOTANY Perennial herb (to 1.2 m); rhizomes and roots aromatic; leaves pinnate; flowers white.
CHEMISTRY Valepotriates (unusual iridoid derivatives, 0.5–2%): valtrate and didrovaltrate; cyclopentane sesquiterpenoids: **valerenic acid**; essential oil (mainly bornyl acetate).
PHARMACOLOGY Valerenic acid and the essential oil are sedative, valtrate is thymoleptic and didrovaltrate is tranquillising. Clinical studies indicate efficacy against minor nervous conditions.
TOXICOLOGY Valepotriates are potentially mutagenic, so that infusions are safer than tinctures.

Valeriana officinalis L. (Valerianaceae); *valériane officinale* (French); *Gemeiner Baldrian, Arzneibaldrian* (German); *valeriana* (Italian)

Vanilla planifolia
vanilla

CLASSIFICATION TM: Central and America, Europe.
USES & PROPERTIES Cured vanill ("beans") were originally used as flavou dient of chocolate, with the reputation of aphrodisiac properties and inducing a fe well-being. Vanilla is widely used in the fo beverage industries and also to mask unp odours and tastes in medicines and cough s
ORIGIN Central America (mainly Mexico
BOTANY Climber (vine) (to 3 m); leaves fleshy; flower greenish yellow; fruit a capsu thousands of minute black seeds.
CHEMISTRY **Vanillin** (2%) is the main f compound, enzymatically released from a g side (vanilloside) when the fruits are cured.
PHARMACOLOGY Vanillin has demons antimutagenic, chemopreventative, antimicr antioxidant and anti-inflammatory activiti shows potential for use against sickle-cell ana and inflammatory bowel disease.
TOXICOLOGY Allergic reactions (but very r

Vanilla planifolia Andr. (Orchidaceae); *vanille* (French); *Echte Vanille* (German); *vaniglia* (Italian); *vainilla* (Spanish)

Veratrum album
white hellebore • false hellebore

CLASSIFICATION Cell toxin, neurotoxin, extremely hazardous (Ia). TM: Europe, Asia.

USES & PROPERTIES All plant parts of this and related species are extremely poisonous and have been used since ancient times for murder and suicide, but also to treat gout, rheumatism and neuralgic pains. It was once an ingredient of sneezing powders and was used to kill lice, insects and rats.

ORIGIN Europe and Asia. Grown in gardens.

BOTANY Perennial herb; leaves broad, whorled, pleated; flowers green and white.

CHEMISTRY Steroidal (triterpenoid) alkaloids: **protoveratrine B** is a main compound, together with protoveratrine A, germerine and cyclopamine. The original source of resveratrol (hence the name).

PHARMACOLOGY The alkaloids are very toxic (they affect Na^+-channels), with mutagenic, emetic, analgesic and antirheumatic activities. They are lipophilic (easily absorbed, even through the skin).

TOXICOLOGY The lethal dose in humans is 20 mg of protoveratrine B (or 1–2 g of the dry rhizome). Cyclopamine causes malformations in animals.

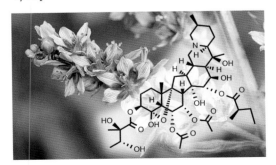

Veratrum album L. (Melanthiaceae); *vérâtre blanc* (French); *Weißer Germer* (German); *veratro bianco* (Italian); *vedegambre, eléboro blanco* (Spanish)

Verbascum phlomoides
mullein • orange mullein

CLASSIFICATION TM: Europe. Pharm., Comm. E+ (flowers only), PhEur8, HMPC.

USES & PROPERTIES Dried flowers (petals with stamens; *Verbasci flos*) are used against catarrh of the respiratory tract (cough, bronchitis and influenza). It is a traditional diuretic and diaphoretic, taken as tea (1 g in 150 ml water), 3–4 × per day. Extracts are used in cough syrups and ointments to treat sores, boils, earache and haemorrhoids.

ORIGIN Europe, western Asia and North Africa. *V. densiflorum* and *V. thapsus* are also used.

BOTANY Biennial herb; leaves in a robust rosette; flowers yellow, in a large panicle (to 2 m).

CHEMISTRY Triterpene saponins (including **verbascosaponin**); mucilage (3%); iridoid glucosides (aucubin, catalpol); flavonoids (kaempferol, rutin).

PHARMACOLOGY Secretolytic and expectorant activity (saponins); demulcent and soothing effects on skin and mucosa (mucilage); weak diuretic activity (flavonoids); anti-inflammatory properties (iridoid glucosides).

TOXICOLOGY No side effects are known.

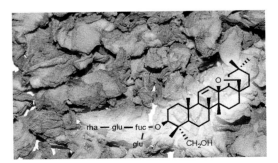

Verbascum phlomoides L. (Scrophulariaceae); *molène faux-phlomis* (French); *Filz-Königskerze* (German); *barbarasco* (Italian)

Verbena officinalis
vervain

CLASSIFICATION TM: Europe, Asia. Pharm., Comm.E–.

USES & PROPERTIES Aboveground parts, collected while flowering (*Verbenae herba*) are traditionally used as diuretic and bitter tonics to treat fever, chronic bronchitis and rheumatism. Vervain herb is externally applied to slow-healing wounds. The recommended dose is 1.5 g of the dry herb, prepared as tea (3–4 g per day).

ORIGIN North temperate region (Europe, Asia, North Africa and North America).

BOTANY Perennial herb (to 1 m); leaves widely spaced, opposite; flowers small, lilac, in spikes.

CHEMISTRY Iridoid glucosides (0.2–0.5%): mainly **verbenalin** (= cornin) and hastatoside. Caffeic acid derivatives: verbascoside.

PHARMACOLOGY The herb is a traditional diuretic, expectorant, galactogogue and antirheumatic but there is limited data. Verbenalin is thought to have antitussive and sleep-promoting activity. The iridoids are anti-inflammatory and analgesic.

TOXICOLOGY There are no known side effects.

Verbena officinalis L. (Verbenaceae); *ma bian cao* (Chinese); *verveine officinelle* (French); *Eisenkraut* (German); *verbena commune* (Italian)

Vinca minor
periwinkle • lesser periwinkle

CLASSIFICATION Cell toxin (II). TM: Europe (now obsolete). MM: source of alkaloids.

USES & PROPERTIES Leaves (*Vincae minoris folium*) were once used in traditional medicine but are now a source of raw material for alkaloid extraction. The alkaloids are given orally (as prescription medicine, 40–60 mg per day) to stimulate cerebral circulation and to treat senility, memory loss and other cerebrovascular disorders.

ORIGIN Europe. A popular garden plant.

BOTANY Small, herbaceous perennial with milky latex; stems trailing; leaves opposite; flowers blue.

CHEMISTRY Numerous monoterpene indole alkaloids (1–4%), mainly **vincamine** (10%) which is nowadays synthesised from tabersonine (a seed alkaloid from African *Voacanga africana*).

PHARMACOLOGY Human studies showed that vincamine reduces blood pressure and stimulates blood flow to the brain.

TOXICOLOGY The herb is potentially toxic and is not suitable for self-medication. Side effects include leukocytopaenia and lymphocytopaenia.

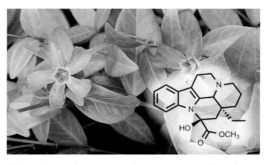

Vinca minor L. (Apocynaceae); *petite pervenche* (French); *Kleines Immergrün* (German); *pervinca minore* (Italian); *hierba doncella* (Spanish)

Viola tricolor
wild pansy • heartsease

CLASSIFICATION TM: Europe. Comm. E+, ESCOP Suppl., PhEur8, HMPC.

USES & PROPERTIES Dried aboveground parts (*Violae tricoloris herba*) are used in supportive treatment of mild seborrhoeic skin conditions (scaly, flaky, itchy and red skin), eczema, pruritus, acne and impetigo. Tea made from 1.5–3 g is taken 3 × per day (4 g in 150 ml for external use). Included in commercial antitussives, cholagogues, roborants, tonics and dermatological remedies.

ORIGIN Europe and Asia. Also cultivated.

BOTANY Annual or short-lived perennial herb; leaves stipulate; flowers typically three-coloured.

CHEMISTRY Salicylic acid derivatives (**methyl salicylate** and violutoside); organic acids, flavonoids, anthocyanins, saponins, tannins and mucilage.

PHARMACOLOGY Methyl salicylate, saponins, flavonoids and mucilage may contribute to the antimicrobial and anti-inflammatory effects. Traditional uses also indicate expectorant, diaphoretic, antirheumatic and diuretic activities.

TOXICOLOGY There are no safety concerns.

Viola tricolor L. (Violaceae); *pensée sauvage* (French); *Feldstiefmütterchen* (German); *viola del pensiero* (Italian)

Viscum album
mistletoe • European mistletoe

CLASSIFICATION Cell toxin (II). TM: Europe. Pharm., Comm.E+. MM: source of extracts for injection; clinical studies+.

USES & PROPERTIES The fresh or dried herb (*Visci herba*) or leaves only (*Visci folium*) are a traditional adjuvant in treating hypertension, vertigo and cephalic congestion. An infusion of 2.5 g is taken 1–2 × per day. This use should be distinguished from parenteral use: special extracts are injected to treat degenerative inflammation of joints and as palliative therapy for malignant tumours.

ORIGIN Europe and Asia.

BOTANY Semi-parasitic woody shrub; leaves leathery, yellowish; flowers small; fruit a white drupe.

CHEMISTRY Numerous lectins (0.1%); polypeptides (called viscotoxins); polysaccharides.

PHARMACOLOGY When injected: cytostatic and cytotoxic at high doses, with non-specific immune stimulation at low doses (ascribed to the lectins and partly also to the polysaccharides).

TOXICOLOGY Not suitable for self-medication. Necrosis and other serious side effects may occur.

Viscum album L. [Santalaceae (formerly Viscaceae)]; *gui blanc* (French); *Mistel* (German); *vischio* (Italian); *muérdago* (Spanish)

Visnaga daucoides
visnaga • khella

CLASSIFICATION TM: Europe, Africa. Pharm., Comm.E– (withdrawn in 1994), WHO 3.

USES & PROPERTIES The ripe fruits (*Ammi visnagae fructus*) or more often standardized extracts are used as preventive treatment of asthma, spastic bronchitis and *angina pectoris*. Infusions of the fruits (0.5 g) are rarely used in traditional medicine for colic and as diuretic to treat kidney ailments.

ORIGIN Mediterranean region.

BOTANY Erect annual herb (to 1.5 m); leaves feathery; flowers small; fruit a small dry schizocarp, borne on persistent rays. The species is usually included in the genus *Ammi* (as *A. visnaga*).

CHEMISTRY Furanocoumarins (khellin, visnagin) and pyranocoumarins (**visnadin**, samidin).

PHARMACOLOGY Visnadin is a strong vasodilator; khellin and visnagin are antispasmodic. Extracts act as muscle relaxants with antispasmodic, vasodilatory and anti-asthmatic activity.

TOXICOLOGY Visnaga is no longer considered to be safe. Side effects include pseudo-allergic reactions, reversible liver ailments and insomnia.

Visnaga daucoides Gaertn. or **Ammi visnaga** (L.) Lam. (Apiaceae); *khella* (Arabian); *herbe aux cure-dents* (French); *Khellakraut, Bischofskraut* (German); *visnaga* (Italian)

Vitex agnus-castus
chaste tree

CLASSIFICATION TM: Europe. Comm.E+, WHO 4, PhEur8, HMPC, clinical studies+.

USES & PROPERTIES Ripe, dried fruits (chaste tree fruit; *Agni casti fructus*) are traditionally used as an anaphrodisiac (hence the name "monk's pepper"). Nowadays they are used for gynaecological, menstrual and menopausal disorders. In homoeopathy, for depression, impotence and hypogalactia (inadequate lactation). The daily dose is up to 3 g.

ORIGIN Europe and Asia (Mediterranean).

BOTANY Shrub (to 5 m); leaves digitate; flowers lilac, blue (white); fruit a small, 4-seeded berry.

CHEMISTRY Diterpenes (**vitexilactone**, rotundifurane); iridoid glucosides (casticin, penduletin); flavonoids; essential oil (1,8-cineole, limonene).

PHARMACOLOGY The diterpenes indirectly inhibit prolactin secretion. Clinical studies support the use of standardised extracts for symptomatic relief of premenstrual stress syndrome, dysmenorrhoea, *corpus luteum* deficiency and mastalgia.

TOXICOLOGY Non-toxic but some side effects have been reported, such as skin rashes.

Vitex agnus-castus L. [Lamiaceae (formerly Verbenaceae)]; *gattilier agneau-chaste* (French); *Mönchspfeffer* (German); *agnocasto* (Italian)

Vitis vinifera
grape vine

CLASSIFICATION Cell toxin, mind-altering (III). TM: Europe. ESCOP Suppl., HMPC. DS: grape seed oil, resveratrol.

USES & PROPERTIES Ripe seeds are extracted for grape seed oil, used as antioxidant and venotonic. Leaves have been used against diarrhoea, poor circulation and bleeding. Skins of red grapes are a source of resveratrol. Red wine has health benefits when used in moderation (the French paradox).

ORIGIN Mediterranean region. Widely cultivated.

BOTANY Deciduous woody climber; stems with tendrils; leaves lobed; flowers small; fruit a berry.

CHEMISTRY Oligomeric proanthocyanidins (pycnogenols) in seeds (80–85%); glucose (20%) and alcohol (to 14%) when fermented; **resveratrol** (to 4%); anthocyanins (cyanidin and petunidin).

PHARMACOLOGY Seed oil and anthocyanins are antioxidant (free-radical scavenging), antimutagenic and reduce capillary fragility. Resveratrol is active *in vitro* (but clinical evidence is lacking).

TOXICOLOGY The lethal dose of ethanol is 5–8 g/kg body weight in adults and 3 g/kg in children.

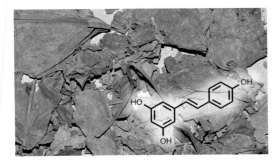

Vitis vinifera L. (Vitaceae); *vigne* (French); *Weinrebe* (German); *vite* (Italian)

Warburgia salutaris
pepperbark tree

CLASSIFICATION TM: Africa. AHP.

USES & PROPERTIES The stem bark (or nowadays also the leaves) are used as a general tonic and panacea to treat coughs, colds, bronchitis and oral thrush (traditionally for headache, influenza, rheumatism, malaria, venereal diseases, chest complaints, gastric ulcers and toothache). Leaves and bark have a sharp peppery taste. Traditionally, cold water infusions were taken as expectorants or powdered bark smoked to treat coughs and colds. Tablets (200 mg of leaf powder) are available.

ORIGIN Africa. Trees have been over-exploited for bark but are grown on a small commercial scale.

BOTANY Evergreen tree (to 10 m); leaves glossy; flowers green; fruit a green to black berry.

CHEMISTRY Several drimane sesquiterpenoids (**warburganal**, polygodial); also mannitol.

PHARMACOLOGY Warburganal and other reactive dialdehydes have demonstrated antibacterial and anti-ulcer activities. Mannitol is diuretic and has been used as a sweetener for diabetics.

TOXICOLOGY No serious side effects are known.

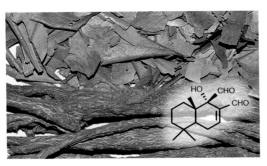

Warburgia salutaris (Bertol.f.) Chiov. (Canellaceae); *isibhaha* (Zulu); *warburgia* (French); *Warburgia, Pfefferrindenbaum* (German); *warburgia* (Italian)

Withania somnifera
winter cherry • ashwagandha

CLASSIFICATION Cell toxin, mind-altering (III). TM: Asia (India), Africa. Pharm., WHO 4, HMPC, clinical studies+.

USES & PROPERTIES The roots (often called "Indian ginseng") are an important general tonic and adaptogen in Ayurvedic medicine, used for a wide range of ailments, including stress and fatigue. The leaves are used topically for wound healing.

ORIGIN Africa, southeastern Europe and Asia.

BOTANY Shrublet (to 1 m); roots fleshy; leaves velvety; fruit a red berry, enclosed in a papery calyx.

CHEMISTRY Free steroids of the ergostane type, the so-called withanolides: **withaferin A** is a major compound. Also alkaloids (withasomnine).

PHARMACOLOGY The withanolides are linked to anti-inflammatory, antibiotic, immunomodulating, cytotoxic, antitumour and cholesterol-lowering activities. The alkaloids are sedative and hypnotic. Clinical studies showed efficacy in treating physical and psychological stress.

TOXICOLOGY Existing data indicate very low oral toxicity. No serious side effects are known.

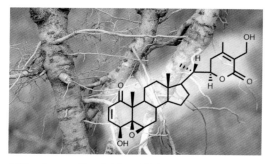

Withania somnifera (L.) Dunal (Solanaceae); *withania* (French); *Schlafbeere, Withania* (German); *ashwagandha* (Hindi); *witania, ginseng indiano* (Italian)

Xysmalobium undulatum
uzara • milk bush

CLASSIFICATION Cell toxin (Ib). TM: Africa (South Africa), Comm.E+.

USES & PROPERTIES Dried roots (*Uzarae radix*) are used mainly to treat non-specific acute diarrhoea. An initial dose of 1 g (75 mg of total glycosides) is followed by 45–90 mg of glycosides per day. Many traditional uses have been recorded, including diarrhoea, colic, stomach cramps, headache, dysmenorrhoea and as diuretic for oedema. Powdered root is applied to treat wounds.

ORIGIN Southern Africa. Cultivated since 1904.

BOTANY Deciduous perennial herb with milky latex; leaves opposite; flowers yellowish brown; fruit an inflated capsule; seeds with long hairs.

CHEMISTRY Cardiac glycosides: **uzarin** (a glycoside of uzarigenin) is the main compound.

PHARMACOLOGY Uzarin stops diarrhoea by a spasmolytic effect on visceral smooth muscles, thus inhibiting intestinal mobility.

TOXICOLOGY Uzarin-type glycosides have a low oral toxicity but can be lethal when injected. Persistent diarrhoea requires medical intervention.

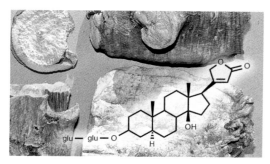

Xysmalobium undulatum R. Br. [Apocynaceae formerly Asclepiadaceae)]; *ishongwe* (Zulu); *uzara* (French); *Uzara* (German); *uzara* (Italian)

Zingiber officinale
ginger

CLASSIFICATION TM: Asia, Europe. Comm.E+, ESCOP 1, WHO 1, PhEur8, HMPC., clinical trials+. DS: anti-emetic.
USES & PROPERTIES The fresh or dried rhizome (*Zingiberis rhizoma*) is used to treat post-operative nausea and travel sickness. Daily dose: 2–4 g. Widely used (e.g. Ayurveda, TCM) for many ailments, including nausea, colic, stomach pain, fever and coughs. Ginger is an important spice.
ORIGIN Asia. Probably an ancient Indian cultigen.
BOTANY Perennial herb (to 1 m) with branched rhizomes; leaves large; flowers yellow and purple.
CHEMISTRY Diterpene lactones (galanolactone); sesquiterpenes (zingiberene, curcumene); pungent gingerols: gingerol (= **6-gingerol**), the main compound; essential oil (camphene, β-phellandrene).
PHARMACOLOGY Clinical studies support the anti-emetic use. Also: anti-inflammatory, anti-microbial, antiparasitic, hypoglycaemic, cholesterol-lowering, immunomodulating, carminative, cholagogic, anti-ulcerogenic, antispasmodic and antioxidant.
TOXICOLOGY Ginger is edible and safe to use.

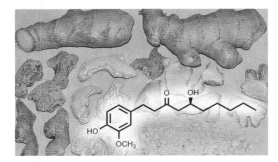

Zingiber officinale Roscoe (Zingiberaceae); *gingembre* (French); *Ingwer* (German); *zenzero* (Italian); *jengibre* (Spanish)

Ziziphus jujuba
Chinese date • jujube tree

CLASSIFICATION Mind-altering. TM: Asia (China). WHO 3. DS: functional food.
USES & PROPERTIES The ripe fruit, fresh or dried (*Jujubae fructus*), is a functional food and general tonic in China, eaten to improve health, to gain weight and to treat upper respiratory tract infections and allergies. Seeds are used to treat insomnia and nervous conditions. Chinese dates are a popular snack and health food in many countries.
ORIGIN Europe and Asia. A fruit crop in China.
BOTANY Shrub or small tree (to 6 m); stems spiny; leaves glossy; flowers yellowish; fruit a drupe.
CHEMISTRY Seeds: peptide alkaloids: **frangufoline** (= sanjoinine A) is the main compound. Fruit pulp: triterpenoid saponins, flavonoids and mucilage.
PHARMACOLOGY The cyclopeptide alkaloids (especially frangufoline) have strong sedative and hypnotic effects. Mucilage is demulcent while the saponins are secretolytic. Fruits seem to act as a tonic by strengthening the immune system, increasing endurance and improving liver function.
TOXICOLOGY Fruits are non-toxic and nutritious.

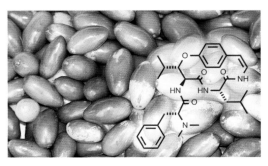

Ziziphus jujuba Mill. [= *Z. vulgaris* Lam.] (Rhamnaceae); *jujubier* (French); *Jujube, Brustbeerbaum* (German); *giuggiolo comune* (Italian)

Glossary of botanical, chemical, medical, pharmacological and toxicological terms

Abiotic: not associated with living organisms
Abortifacient: a substance that causes abortion
Absolute lethal concentration (LC$_{100}$): lowest concentration of a substance that kills 100% of test organisms or species under defined conditions; this value is dependent on the number of individuals used in its assessment
Absorbed dose: amount of a substance absorbed into an organism or into organs and tissues of interest
Absorption: active or passive uptake of a substance through the skin or a mucous membrane; substances enter the bloodstream and are transported to other organs
Accepted risk: probability of suffering disease or injury which is accepted by an individual
Accidental exposure: unintended contact with a substance resulting from an accident
Accumulation: successive additions of a substance to a target organism, or organ, which leads to an increasing amount of the chemical in the organism or organ
ACE: angiotensin converting enzyme; important target for the treatment of cardiac insufficiency
Acetylcholine: a neurotransmitter that binds to nicotinic (nACh-R) or muscarinic (mACh-R) receptors
Acetylcholine esterase inhibitor: a substance that inhibits acetylcholine esterase (AChE) and thus the breakdown of acetylcholine to acetate and choline
Acidosis: pathological condition in which the pH of body fluids is below normal and therefore the pH of blood falls below the reference value of 7.4
Acne: chronic skin condition resulting from the inflammation of sebaceous glands and hair follicles
Acute: a symptom or condition that appears suddenly (and lasts for a short period; minutes or hours)
Acute toxicity test: experimental animal study to determine which adverse effects occur in a short time (usually up to 14 days) after a single dose of a substance or after multiple doses given in up to 24 hours
Adaptogen: a substance with a non-specific action that causes improved resistance to physical and mental stress
Adaptogenic tonic: see Adaptogen
Addiction: physical or psychological dependence on a substance for the sake of relief, comfort, stimulation, or exhilaration; often with craving when the drug is absent
Additive: a substance that is added to a mixture (typically for taste, colour or texture)
Additivity: the effect of a combination of two or more individual substances is equivalent to the sum of the expected individual responses
Adduct: new chemical species AB, formed by direct combination of two separate molecular entities A and B
Adenocarcinoma: malignant tumour originating in glandular epithelium or forming recognisable glandular structures
Adenoma: an abnormal benign growth of glandular epithelial tissue
Adenylyl cyclase: enzyme of signal transduction; catalyses the formation of the second messenger cAMP from ATP
Adjuvant: a substance added to a mixture to modify the activity of the active ingredient in a predictable way
Administered dose: a defined quantity given orally or parenterally

Administration: application of a known amount of a chemical in a reproducible manner and by a defined route
Adrenalin: the hormone that binds to adrenergic receptors; causes the "fight or flight" response
Adrenergic (sympathomimetic): a substance that binds to adrenergic neuroreceptors and produces an effect similar to normal impulses (caused by adrenaline, noradrenaline) of the sympathetic nervous system; antagonists are sympatholytics
Adsorption: enrichment of substances on surfaces
Adulterant: an undesirable ingredient found in a commercial product
Adulteration: non-allowed substitution of a substance or materials in a drug or food by another, usually being inactive or toxic
Aesthenia (astenia): diminishing strength and energy; weakness
Aetheroleum: essential (volatile) oil
Aetiology: science dealing with the cause or origin of disease
Aglycone: the non-sugar part of a glycoside (after removal of the sugar part)
Agonist: substance that binds and activates a cellular receptor
AIDS: acquired immunodeficiency syndrome, a condition (weakened immune system) caused by HIV (a retrovirus)
Albuminuria: presence of albumin from blood plasma in the urine
Alcoholic extract: soluble fraction of plant material obtained after extraction with ethanol
Alkaloid: chemical substance containing nitrogen as part of a heterocyclic ring structure; often highly toxic or mind-altering
Alkylating agent: reactive secondary metabolite which introduces an alkyl substituent into DNA, proteins or other molecules; DNA alkylation can lead to mutations and cancer
Alkylation: reactive secondary metabolites forming covalent bonds with DNA and proteins
Allergen: an antigenic substance that triggers an allergic reaction (hypersensitivity)
Allergy: a hypersensitivity to allergens (often pollen or reactive secondary metabolite) that causes rhinitis, urticaria, asthma and contact dermatitis
Allopathy: a medicine system using substances that cause an effect different to those caused by the condition under treatment
Alopecia: loss of hair
Alzheimer's disease: see Dementia
Amarum: a bitter substance that stimulates the secretion of digestive juices
Ames test: *in vitro* test for mutagenicity using mutant strains of the bacterium *Salmonella typhimurium*. The test can be carried out in the presence of a given microsomal fraction (S-9) from rat liver to allow metabolic transformation of mutagen precursors to active derivatives
Amino acid: chemical substances that form the building blocks of proteins ("proteinogenic amino acids")
Amoebiasis: a (sub-)tropical protozoan infection with *Entamoeba histolytica*

Amoebicidal: a substance that kills amoebae

Anaemia (anemia): reduced number of red blood cells in the blood, often causing pallor and fatigue

Anaesthetic: a substance that causes localised or general loss of feeling or sensation; general anaesthetics produce loss of consciousness; local anaesthetics render a specific area insensible to pain

Analeptic: a substance that stimulates the central nervous system

Analgesic: a substance that relieves pain without loss of conscience

Anaphrodisiac: a substance that reduces sexual desire

Anaphylactic shock: a severe, life-threatening form of a general allergic reaction to an antigen or hapten to which a person has previously been sensitised

Aneuploid: missing or extra chromosomes or parts of chromosomes

Angina pectoris: severe pain in the chest

Anorectic: a substance that suppresses appetite

Anoxia: total absence of O_2, refers sometimes to a decreased oxygen supply in tissues

Antagonist: inhibitor at cellular receptors; blocks the activity of an endogenous ligand; reverses or reduces the effect modulated by an agonist

Anthelmintic: a substance that kills or expels intestinal worms

Anthraquinones: secondary metabolites with an anthracene skeleton; anthrones show strong laxative effects

Anthroposophic: a philosophy that links health to internal life force and energy

Anti-arrhythmic: a substance that counteracts irregular heartbeat

Anti-asthmatic: a substance that alleviates the spasms of asthma

Antibacterial: a substance that kills or inhibits the growth of bacteria

Antibiotic: a substance that kills or inhibits the growth of microorganisms

Antibody: specific protein produced by the immune system (an immunoglobulin molecule), which can bind specifically to an antigen or hapten which induced its synthesis

Anticholinergic: a substance that blocks the parasympathetic nerve impulse

Anticoagulant: a substance that prevents blood from clotting

Anticonvulsant: a substance that prevents or relieves convulsions

Antidepressant: a substance that alleviates depression

Antidiabetic: a substance that prevents or alleviates diabetes

Antidiuretic: a substance that prevents or slows urine formation

Antidote: a substance that counteracts the effect of a potentially toxic substance

Anti-emetic: a substance that prevents vomiting

Antifungal: a substance that kills or inhibits the growth of fungi

Antigen: substance which induces the immune system to produce a specific antibody or specific cells

Antihistamine: a substance that improves allergic symptoms by blocking the action of histamine

Antihydrotic: a substance that reduces perspiration

Antihypertensive: a substance that reduces high blood pressure

Anti-inflammatory: a substance that causes relief of inflammation

Antimetabolite: a substance structurally similar to a metabolite, which competes with it or replaces it, and thus prevents or reduces its normal function

Antimicrobial: a substance that kills or inhibits the growth of microorganisms

Antimitotic: a substance that prevents or inhibits cell division (mitosis)

Antimycotic: *see* Antifungal

Anti-oedemic: a substance that prevents swelling

Antioxidant: a substance that is able to protect cells or counteract the damage caused by oxidation and free oxygen radicals (reactive oxygen species, ROS)

Antiparasitic: a substance that kills parasites

Antiphlogistic: a substance that prevents inflammation

Antipruritic: a substance that alleviates or prevents itching

Antipyretic: a substance that alleviates fever

Antirheumatic: a substance that relieves the symptoms of rheumatism

Antiseptic: a substance that stops or inhibits infection

Antispasmodic: a substance that reduces muscular spasms and tension

Antitumour: a substance that counteracts tumour formation or tumour growth

Antitussive: a substance that reduces the urge to cough

Anuria: the inability to urinate

Anxiety: symptoms of fear not caused by any danger or threat

Aperitif: a drink that stimulates the appetite

Aphasia: loss or impairment of the power of speech or writing, or of the ability to understand written or spoken language or signs, due to a brain injury, disease or drugs

Aphrodisiac: a substance that increases sexual desire

Apnoea (apnea): cessation of breathing

Apoptosis: programmed cell-death leading to a progressive fragmentation of DNA and disintegration of cells without causing inflammation

Aqueous extract: soluble fraction of plant material obtained after extraction with water

Aril: fleshy edible structure attached to a seed and aiding its dispersal by animals

Aromatherapy: the medicinal use of aroma substances by inhalation, bath, massage, etc.

Aromatic bitter: a partly volatile substance that promotes appetite and digestion by stimulating the secretion of digestive juices

Arrhythmia: any deviation from the normal rhythm of the heartbeat

Arteriosclerosis: accumulation of fatty deposits in the blood vessels causing them to narrow and harden, resulting in heart disease or stroke

Arthritis: inflammation of joints

Asphyxiant: substance that blocks the transport or use of oxygen by living organisms

Asthma: chronic respiratory disease characterised by bronchoconstriction, excessive mucus secretion and oedema of the pulmonary alveoli, resulting in difficulty in breathing out, wheezing and coughing

Astringent: a substance (often tannins) that reacts with proteins in wounds, on the surface of cells or membranes, resulting in a protective layer and causing contraction

Ataxia: loss of muscle coordination leading to unsteady or irregular walking or movement

Atherosclerosis: pathological condition; changes of arterial walls that lead to arteriosclerosis

Atrophy: wasting away of the body or of an organ or tissue

Autonomic nervous system: that part of the nervous system that regulates the heart muscle, smooth muscles and glands; it comprises the sympathetic nervous system and the parasympathetic nervous system

Autopsy: post-mortem examination of the organs and body tissue to determine cause of death

Ayurvedic medicine: traditional medicine in India

Bactericide: substance that kills bacteria

Bacteriostatic: a substance that prevents the multiplication of bacteria

Bacterium: a microorganism consisting of a single cell surrounded by a cell wall; DNA is circular; bacteria do not have internal membrane systems or a nucleus

Base pairing: complexation of the complementary pair of polynucleotide chains of nucleic acids by means of hydrogen bonds between complementary purine and pyrimidine bases, adenine (A) with thymine (T) or uracil, cytosine (C) with guanine (G)

Benign: (1) not cancerous, not malignant since tumour does not form metastases and has still positional control. (2) disease without persisting harmful effects

Benign prostatic hyperplasia: a non-cancerous enlargement of the prostate that may interfere with urination

Benzodiazepine receptor: binding site for benzodiazepines at the GABA receptor; target for several sedatives, tranquillisers and alcohol

Berry: fleshy fruit with many seeds (*see* Drupe)

Beta-carotene: an orange plant pigment that is converted in the body to vitamin A

Bidesmosidic saponins: saponins with two sugar chains

Bile: a bitter fluid excreted by the liver via the gall bladder that helps to digest fats

Biliary dyskinesia: inability to secrete bile

Bilirubin: orange-yellow pigment, a breakdown product of haem-containing proteins (haemoglobin, myoglobin, cytochromes), it is excreted in the bile by the liver

Bioaccumulation: when harmful substances enter the ecosystem, some of them move through the food chain, by one organism eating another. In the end, substances accumulate, sometimes in high concentration in the top consumer or predator (often humans)

Bioactivation: any metabolic conversion of a xenobiotic to a more toxic derivative (e.g. pyrrolizidine alkaloids)

Bioassay: an experiment in which test organisms are exposed to varying concentrations of a substance. The response of the test organism is determined as a function of experimental conditions

Bioavailability: amount of a substance that is available for pharmacological or toxicological response after absorption

Biomembrane: permeation barrier around every cell or cellular compartments consisting of phospholipids, cholesterol and membrane proteins

Biopsy: excision of a small piece of living tissue for microscopic or biochemical examination and diagnosis

Bitter: a substance that stimulates the secretion of digestive juices

Bitter tonic: a substance that promotes appetite and digestion by stimulating the secretion of digestive juices

Bitterness value: that concentration at which a bitter substance can still be tasted (bitterness value of 100 000: when one part of the substance in 100 000 parts of water still tastes bitter)

Blood–brain barrier: blood vessels of the brain are covered with especially tight endothelial tissues, so that only selective substances can enter the brain

Blood purifier: a substance that causes the removal of impurities from the bloodstream (outdated term)

Bract: leaf-like structure inserted at the base of a flower or flower stalk

Bradycardia: pulse under 60 beats per minute

Bronchitis: inflammation of the mucous membranes of the bronchial tubes

Bronchodilatory: a substance that expands air passage through the bronchi and reduces bronchial spasm

Bruise: a non-bleeding injury to the skin

Bulbus: dried bulbs

Cachexia: weight loss due to chronic illness or prolonged emotional stress

Calyx: outer circle of leaf-like structures surrounding a flower (usually green)

Cancer: various types of malignant cells that multiply out of control

Candidiasis: infection with the fungus *Candida albicans*

Carcinogen: a substance that may cause cancer

Carcinogenicity: a multistage process leading to abnormal cell growth and cell differentiation; during initiation cells undergo mutations, during promotion mutated cells are stimulated (e.g. by co-carcinogens) to progress to cancer

Carcinoma: malignant growth of epithelial cells

Cardiac glycoside: a steroidal glycoside that inhibits NA^+, K^+ ATPase and thereby indirectly increases the strength or rhythm of the heartbeat

Cardiotonic: a substance that has a strengthening or regulating effect on the heart

Cardiotoxic: harmful to heart cells

Carminative: a substance that reduces flatulence

Catabolism: process of breakdown of complex molecules into simpler ones, often providing biologically available energy in form of ATP

Catalyst: a compound that speeds up the rate of a reaction

Catarrh: inflammation of mucous membranes

Catechol-*O*-methyltransferase (COMT): enzyme which inactivates neurotransmitters with a phenolic OH group (dopamine, noradrenaline, serotonin) through methylation

Cathartic: laxative, purgative

Catkin: cluster of small naked (apetalous) flowers

Chemotherapy: treatment of cancer with cytotoxic substances

Chiral: if the mirror image of a substance is not superimposable, it is called chiral

Cholagogue: a substance that stimulates the flow of bile from the gall bladder

Cholekinetic: a substance that stimulates the release of bile by contraction of the gall bladder and bile ducts

Choleretic: a substance that stimulates the liver to produce bile

Cholesterol: the most common steroid (fat-like material) found in the human body; important for membrane fluidity and as a precursor for steroid hormones; high cholesterol levels are associated with an increased risk of coronary diseases

Cholinesterase inhibitor: a substance that inhibits the action of cholinesterase (AChE); AChE catalyses the hydrolysis of acetylcholine (ACh) esters: a cholinesterase inhibitor causes hyperactivity in parasympathetic nerves

Chromosomal aberration: abnormal chromosome number or structure

Chronic: occurring over a long period of time (>1 year)

Chronic ailment: a condition that extends over a long period

Chronic exposure: continued exposure over an extended period of time

Chronotoxicology: science of the influence of biological rhythms on the toxicity of substances
Cirrhosis: liver disease defined by increased fibrous tissue, with loss of functional liver cells, and increased resistance to blood flow through the liver portal
Clinical trials: the development of new drugs consists of four phases: 1. preclinical studies, 2. clinical studies phase I, 3. clinical studies phase II and 4. clinical studies phase III
CNS: central nervous system
Co-carcinogen: a substance that amplifies the effect of a carcinogen or promotes tumour development
Colic: abdominal pains, caused by muscle contraction of an abdominal organ, accompanied by nausea, vomiting and perspiration
Commission E: recommendations of a group of German experts regarding the usefulness and efficacy of plant drugs
Concentration: the amount of a given substance in a given volume of air or liquid
Concentration–response curve: graph of the relation between exposure concentration and the magnitude of the resultant biological change
Condyloma: warts of the genital-anal region (caused by viruses of the Papilloma group)
Conjugate: derivative of a substance formed by its combination with chemicals such as acetic acid, glucuronic acid, glutathione, glycine, sulphuric acid, etc.
Conjunctiva: the mucous membranes of the eyes and eyelids
Conjunctivitis: inflammation of conjunctiva
Constipation: lack of bowel movement leading to prolonged passage times of faeces
Contact dermatitis: inflammatory condition of the skin resulting from dermal exposure to an allergen or an irritating chemical
Contaminant: any kind of adverse substance that contaminates water, air or food
Contraindication: condition that makes some particular treatment improper or undesirable
Corpus luteum: endocrine body in the ovary that secretes oestrogen and progesterone
Cortex: dried bark
Covalent bond: a bond created between two atoms when they share electrons
Crohn's disease: chronic inflammation of the intestinal tract
Cumulative effect: mutually enhancing effects of repeated doses of a harmful substance
Cutaneous: relating to the skin
Cyanogenic glucoside: secondary metabolite that is activated upon wounding, releasing the toxin HCN
Cyanosis: bluish coloration, especially of the skin and mucous membranes and fingernail beds; occurs when oxygenation is deficient and reduced haemoglobin is abundant in the blood vessels
Cyclooxygenase: key enzyme of prostaglandin biosynthesis converting arachidonic acid into prostaglandins
Cystitis: inflammation of the bladder
Cytochrome P-450: important haemoprotein which has the task to hydroxylate many endogenous and exogenous substrates (which are later conjugated and excreted). The term includes a large number of isoenzymes which are coded for by a superfamily of genes
Cytoplasm: basic compartment of the cell (surrounded by the plasma membrane) in which nucleus, endoplasmic reticulum, mitochondria and other organelles are imbedded

Cytostatic: a substance that slows down cell growth and multiplication
Cytotoxic: a substance that is toxic to cells, i.e. damages cell structure or function
Decoction: watery extract obtained by boiling
Decongestant: a substance that removes mucus from the respiratory system and opens the air passages so that breathing becomes easier
Dementia: loss of individually acquired mental skills; Alzheimer's disease is a severe form of dementia
Demulcent: a substance that soothes the mucous membranes (sometimes the term is restricted to internal membranes; *see* Emollient)
Dependence: psychic craving for a drug or other substance which may or may not be accompanied by a physical dependency
Depression: psychic disturbance, often associated with low concentrations of dopamine and noradrenaline
Dermal: referring to the skin
Dermal absorption: absorption through the skin
Dermatitis: inflammation of skin (e.g. by contact dermatitis)
Detergent: a substance capable of dissolving lipids
Detoxification: biochemical modification which makes a toxic molecule less toxic
Developmental toxicity: adverse effects on the embryo or growing foetus
Diabetes mellitus: abnormally high blood sugar levels caused by lack of insulin
Diaphoretic: a substance that increases sweating (profuse perspiration)
Diarrhoea: abnormally frequent discharge of watery stool (more than three times per day)
Dietary supplement: a substance that is marketed and sold as a "healthy" food item but not as a therapeutic agent
Diffusion: the process by which molecules migrate through a medium and spread out evenly
Digitate: compound leaf with leaflets arising at the same point (like fingers on a hand)
Disulphide bridge: a bond between two SH-groups, e.g. in a protein
Diuresis: discharge of urine
Diuretic (aquaretic): a substance that increases the volume of urine
DNA: deoxyribonucleic acid, the biomolecule in cells that stores the genetic information; composed of two complementary nucleic acid strands bonded by G-C and A-T pairs
Doctrine of signatures: old concept of traditional medicine assuming that the form or colour of a plant could indicate its medicinal application
Dosage: dose expressed as a function of the organism being dosed and time, for example mg/kg body weight/day
Dose: the amount of a substance to which a person or test organism is exposed; the effective dose depends on body weight
Dose-response curve: graph of the relationship between dose and the degree of changes produced
Dropsy: outdated term for oedema
Drug: term for a therapeutic agent, but also commonly employed for abused substances
Drug-resistance: having a (often acquired) resistance against a drug, by developing modified targets, increasing the degradation of an active compound or by exporting it out of a cell
Drupe: fleshy fruit with a single seed

Dysentery: inflammation of the colon; often caused by bacteria (shigellosis) or viruses, accompanied by pain and severe diarrhoea
Dysmenorrhoea: abnormal or painful menstruation
Dyspepsia: indigestion
Dysplasia: abnormal development of an organ or tissue
Dyspnoea: difficult breathing
Dysuria: painful urination
Eczema: acute or chronic inflammation of the skin with redness, itching, papules, vesicles, pustules, scales, crusts or scabs
Effective concentration (EC): EC_{50} is the concentration that causes 50% of maximal response
Effective dose (ED): ED_{50} is the dose that causes 50% of maximal response
Electronegative: atoms that draw electrons of a bond toward itself (e.g. oxygen)
Elixir: a nonspecific term generally applied to a liquid alcoholic preparation, emulsion or suspension
Embryotoxicity: any toxic effect on the conceptus as a result of prenatal exposure during the embryonic stages of development, including malformations, malfunctions, altered growth, prenatal death and altered postnatal function
Emesis: vomiting
Emetic: a substance causing vomiting
Emollient: a substance that soothes and softens the skin
Endocrine: pertaining to hormones or to the glands that secrete hormones directly into the bloodstream
Endoplasmic reticulum: endomembrane system in which proteins are modified post-translationally
Endorphins: peptides made by the body with similar activities as morphine
Endothelia: layer of cells lining the inner surface of blood and lymphatic vessels
Enteritis: inflammation of the intestines
Entheogen: an intoxicating or hallucinogenic substance that is taken to bring on a spiritual experience
Enzyme: protein that catalyses a chemical reaction, e.g. the hydrolysis of acetylcholine
Epidemiology: science that studies the occurrence and causes of health conditions in human populations; scientists try to find out whether a factor (e.g., nutrient, contaminant) is associated with a given health effect
Epigastric: referring to the upper-middle region of the abdomen
Epilepsy: chronic brain condition characterised by seizures and loss of consciousness
Epileptiform: occurring in severe or sudden spasms, as in convulsion or epilepsy
Epithelia: cell layer covering the internal and external surfaces of the body
Epitope: any part of a molecule that carries an antigenic determinant
Ergot: a fungus (*Claviceps purpurea*) that infects grasses (especially rye) and produces pharmacologically active alkaloids
Ergotism: poisoning by eating ergot-infected grain
Erythema: redness of the skin produced by congestion of the capillaries
Essential oil (=volatile oil): mixture of volatile terpenoids and phenylpropanoids responsible for the taste and smell of many plants, especially spices
Ethnobotany: study of how different human cultures use plants for medicinal and other purposes
Excretion: elimination of chemicals or drugs from the body, mainly through the kidney and the gut. Volatile compounds may be eliminated by exhalation. In the GI tract elimination may take place via the bile, the shedding of intestinal cells and transport through the intestinal mucosa
Expectorant: a substance that increases mucous secretion or its expulsion from the lungs; distinction between secretolytics and secretomotorics
Exposure: contact with a substance by swallowing, breathing, or directly through skin or eye; we distinguish between short-term and long-term exposure
Extract: a concentrated preparation (semi-liquid, solid or dry powder) of the soluble fraction of plant material
Familial Mediterranean fever: a condition with recurrent attacks of fever and pain
Febrifuge: a substance that reduces fever
Fibrinolytic: the ability of some proteolytic enzymes to dissolve fibrin in blood clots, facilitating wound healing
Febrile: relating to fever
First-pass effect: biotransformation of a chemical in the liver (after absorption from the intestine and before it reaches the systemic circulation)
Flatulence: accumulation of excessive gas in the intestines
Flos: dried flowers
Flu: *see* Influenza
Fluid extract: an alcohol–water extract concentrated to the point where, e.g., 1 ml equals 1 g of the original herb
Fluor albus (=leukorrhoea): white or yellow vaginal discharge
Folium: dried leaves
Food allergy: hypersensitivity reaction to chemicals in the diet to which a person has previously been exposed and sensitised
Forced diuresis: clinical method of stimulating diuresis, with the aim of achieving increased clearance of a toxic substance in urine
Frame-shift mutation: point mutation deleting or inserting one or two nucleotides in a gene, shifting the normal reading frame and causing the formation of functionless proteins
Free radical: an unstable form of oxygen molecule that can damage cells and cellular macromolecules
Fructus: dried fruits
Functional food: a food item with some pharmacological activity in addition to nutritional benefits
Galactogogue: a substance that stimulates milk secretion
Galenical preparations: preparations of herbal drugs, such as tinctures, lotions, extracts etc. (often interpreted as referring to non-surgical medicine)
Gallstone: a solid or semi-solid body in the gall bladder or bile duct
Gallotannin: polyphenol present in many medicinal plants; forms hydrolysable tannins (esters of gallic acid with sugars)
Gargle: a fluid used as throat wash
Gastritis: inflammation of the stomach
Gastroenteritis: inflammation of the gastrointestinal tract, associated with nausea, pain and vomiting
Genetic toxicity: damage to DNA by a mutagen, causing mutations and altered genetic expression (mutagenicity). Non-repaired mutations in somatic cells are inherited to daughter cells whereas mutations in germ cells can reach the next generation
Gingivitis: inflammation of the gums
Glaucoma: an eye disease characterised by increased intra-ocular pressure
GLC: high resolution gas-liquid chromatography (a technique used to analyse volatile chemical compounds and extracts)

Glumes: outer bracts of a grass spikelet
Glucosinolate: secondary metabolite that becomes activated upon wounding of a plant, releasing active isothiocyanate
Glycoprotein: protein that carries sugar groups
Glycoside: a chemical substance that yields at least one simple sugar upon hydrolysis
GMP: good manufacturing practice; a manufacturing system that complies with the highest standards of hygiene, safety and quality
Gout: increased uric acid level in blood and sporadic episodes of acute arthritis
Granulation: new cell layers (in the form of small granular prominences) over capillaries and collagen in a wound
GRAS: abbreviation for "generally regarded as safe", the status given to foods and herbal medicines by the American Food and Drug Administration (FDA)
Gravel: small concretions in the bladder or kidney
Haematoma: local accumulation of clotted blood
Haematuria: blood in the urine
Haemodialysis: removal of toxins from the blood through dialysis, using an artificial kidney (allowing the diffusion of toxins from the blood)
Haemolysis: the disruption of red blood cells and release of haemoglobin in blood
Haemoperfusion: removal of toxins from the blood with the aid of a column of charcoal or adsorbent resin
Haemorrhage: profuse bleeding
Haemorrhagic nephritis: blood in the urine
Haemorrhoids (=piles): painful and swollen anal veins
Haemosorption perfusion: passage of a patient's blood through a set of columns filled with a haemosorbent (activated charcoal, ion-exchange resin, etc.); the purpose of the procedure is to remove a toxic substance from the organism, particularly in an emergency
Haemostatic: a substance that reduces or stops bleeding
Haemostyptic: a substance that reduces or stops bleeding
Hallucinogen: a substance that induces the perception of objects which are not actually present
Hazard: capability of an agent to cause adverse effects
Heartburn: uncomfortable burning sensation in the chest, rising towards the throat (due to the return of stomach acid into the oesophagus)
Hepatitis: inflammation of the liver
Hepatotoxic: toxic to the liver
Herbalist: a person with experience in herbal medicine and/or herbal therapy
Herpes simplex: localised infection on the lips or genitalia caused by the herpes virus
HIV: human immunodeficiency virus that causes AIDS
Hodgkin's disease: a cancer of lymph cells that originates in one lymph node and later spreads to other organs
Homeopathy: a medicine system using minute amounts of substances that cause in a healthy person the same effect (symptoms) than those caused by the condition under treatment
Hormone: a substance released into the bloodstream that affects organ systems elsewhere in the body
HPLC: high performance liquid chromatography (a technique used to analyse chemical compounds and extracts)
Hydrogen bond: an attraction between a hydrogen atom (H) on an electronegative atom (mostly N or O)
Hydrolysis: breaking down a molecule by addition of water; in cells hydrolases (glucosidases, lipases, DNAses) catalyse this reaction
Hydrophilic substance: a substance soluble in water but not in oil
Hydrophobic substance: a substance that is repelled by water, but soluble in lipids
Hyperaemia (hyperemia): abnormal blood accumulation in a localised part of the body
Hyperlipidemia: characterised by enhanced lipid levels in the blood; triglycerides (>160 mg/100 ml) and cholesterol (>260 mg/100 ml)
Hyperplasia: abnormal growth of normal cells in a tissue or organ
Hypersensitivity: allergic reaction of a person to chemicals to which they had been exposed previously
Hypertension (hypertonia): high blood pressure (>140/90 mm Hg)
Hypertonic solution: abnormally high salt levels having a higher osmotic pressure than blood or another body fluid
Hypertrophy: abnormal increase in size of an organ (cell numbers remain constant)
Hypnotic: a substance that induces sleep
Hypoglycaemic: abnormally low level of blood sugar
Hypothermia: low body temperature
Hypotonia/hypotension: low blood pressure (<105/60 mm Hg)
Hypoxia: abnormally low oxygen content or tension in the body
Icterus: jaundice; deposition and retention of bile pigment in the skin
Immune stimulant: a substance capable of improving the immune system
Immunosuppression: reduction of the immune response
In vitro: in the laboratory or test tube
In vivo: in a living animal or human
Incidence rate: measure of the frequency of new events occurring in a population
Inflammation: localised swelling, redness and pain as a result of an infection or injury
Inflorescence: flower cluster
Influenza (flu): an acute and highly contagious disease caused by viruses that infect mucous membranes of the respiratory tract
Ingestion: swallowing (eating or drinking) of chemicals; after ingestion chemicals can be absorbed from the GI tract into the bloodstream and distributed throughout the body
Inhalation: exposure to a substance through breathing it; if taken up from the lungs a substance can enter the bloodstream
Inhibitory concentration (IC): IC_{50} is the concentration that causes 50% inhibition
Inorganic agent: material that consists of elements or inorganic compounds
Inotropic: a substance that stimulates the contraction of muscles, e.g. of the heart
Insomnia: inability to sleep
Insulin: a hormone made in the pancreas that controls the level of glucose in the blood
Intercalation: planar and lipophilic compounds can intercalate between base stacks of DNA; this leads to frame shift mutations (resulting in inactive proteins)
Intoxication: (1) Poisoning with clinical signs and symptoms (2) Drunkenness from ethanol-containing beverages or other compounds affecting the central nervous system
Ion channel: membrane protein that can form water-containing pores so that mineral ions can enter or leave cells

Ionic bond: a bond created when two atoms trade electrons and then attract each other due to their opposing charges (e.g. NH_3^+-groups form ionic bonds to COO^--groups)

Iridoids: a subgroup of monoterpenoids, with iridoid glucosides, secoiridoids and secologanin

Irrigation therapy: rinsing the urinary tract by means of a diuretic substance that increases urine flow

Irritant: a substance causing irritation of the skin, eyes or respiratory system

Isothiocyanate: secondary metabolites released from glucosinolates upon hydrolysis; exhibits strong skin irritating properties

Itch (=pruritus): skin irritation

Jaundice: yellow coloration of skin and mucosa; caused by abnormally high level of bile pigments in the blood

Kampo medicine: traditional Japanese medicine

Lacrimator: substance which can irritate the eyes and cause tear formation

Lactation: production and secretion of milk by female mammary glands

Latency: time period from the first exposure to a substance until the appearance of biological effects

Lavage: irrigation or washing out of the stomach, intestine or the lungs

Laxative: substance that causes evacuation of the intestinal contents

LD_{100}: lethal dose that kills all (100%) of the individuals in a test group

LD_{50}: lethal dose that kills 50% of the individuals in a test group

Lethal: deadly; fatal; causing death

Leukaemia: malignant disease of the blood-forming organs

Leukopenia: low white blood cell count

Leukorrhea: vaginal discharge of white or yellowish fluid

Ligand: substance that binds to a receptor in a specific way like a key in a lock

Lignum: dried wood

Liniment: ointment for topical application

Lipid: a substance soluble in non-polar solvents; insoluble in water

Lipid-lowering: a substance that lowers triglyceride or cholesterol levels in blood

Lipophilic (=hydrophobic): a substance soluble in oil or a non-polar solvent

Lowest-observed-adverse-effect-level (LOAEL): lowest concentration of a substance which causes an adverse effect

Lowest-observed-effect-level (LOEL): lowest concentration or amount of a substance which causes any biological effects

Lysosome: cytoplasmic organelle containing hydrolytic enzymes

Maceration: preparation made by soaking plant material

Malaise: slight feeling of bodily discomfort

Malaria: a parasitic disease caused by *Plasmodium* parasites; it is transmitted by mosquitoes

Malignant: a disease which gets progressively worse and results in death if not treated, or a cancer with uncontrolled growth and metastasis

Mania: emotional disturbance characterised by an expansive and elated state (euphoria), rapid speech, flight of ideas, decreased need for sleep, distractibility, grandiosity, poor judgement and increased motor activity

MAO inhibitor: inhibitor of monoamine oxidase that degrades the neurotransmitters adrenaline, noradrenaline, dopamine and serotonin

Mastitis: inflammation of the breast

Mastodynia: pain in the swollen female breasts

Materia medica: the various materials (from plants, animals or minerals) that are used in medicine (healing)

Maximum tolerable dose (MTD): highest amount of a substance that does not kill test animals (LDo)

MDR: multiple drug resistance; caused by overexpression of *p*-glycoprotein, an important ATP-driven transporter at biomembranes, which pumps out lipophilic xenobiotics

Melanoma: a tumour of skin and mucosa arising from the pigment-producing cells

Menopause: permanent cessation of menstruation caused by decreased production of female sex hormones

Menorrhagia: abnormally severe menstruation

Metabolic syndrome: a dietary ailment associated with obesity and diabetes

Metastasis: movement of cancer cells from one part of the body to another, starting new tumours in other organs

Microtubules: linear tubular structures of higher cells, formed from tubulin dimers; essential for cell division and vesicular transport processes

Micturition: urination

Migraine: recurrent condition of severe pain in the head accompanied by other symptoms (nausea, visual disturbance)

Mineralocorticoid: the steroid of the adrenal cortex (aldosterone) that regulates salt metabolism

Minimum lethal dose (LD_{min}): lowest amount of a substance that may cause death

Miosis: abnormal contraction of the pupil to less than 2 mm

Mitochondria: important compartment of eukaryotic cells; site of the Krebs cycle and respiratory chain (production of ATP); mitochondria have their own DNA, replication, transcription and ribosomes

Mitogen: chemical which induces mitosis and cell proliferation

Mitosis: cell division

Monoamine oxidase (MAO): the enzyme that catalyses the removal of amine groups (e.g. dopamine, noradrenaline)

Monodesmosidic saponins: saponins with one sugar chain

Morbidity: any form of sickness, illness and morbid condition

Mucilage: solution of viscous (slimy) substances (usually polysaccharides) that form a protective layer over inflamed mucosal tissues

Mucolytic: a substance that dissolves mucous, e.g. in the bronchia

Mucosa: mucous tissue layer on the inside of the respiratory or gastrointestinal tract

Mucus: clear, viscose secretion formed by mucous membranes

Multiple sclerosis: disorder of the CNS caused by a destruction of the myelin around axons in the brain and spinal cord that lead to various neurological symptoms

Mutagenic: a substance that induces genetic mutations; resulting in alterations or loss of genes or chromosomes

Myalgia: non-localised muscle pain

Mycotoxin: toxin produced by a fungus

Mydriasis: dilation of the pupil of the eye

Na^+, K^+-ATPase: important ion pump of animal cells; pumps Na^+ out of the cell and K^+ into the cell; is inhibited by cardiac glycosides

Narcotic: a substance that produces insensibility or stupor, combined with a sense of well-being; more specifically an opioid, any natural or synthetic drug that has morphine-like activity

Naturopathy: a holistic system of healing that emphasises the body's inherent power of regaining balance and harmony

Necrosis: death of cells or tissue, usually accompanied by an inflammation

Neoplasm: new and abnormal formation of a tumour by fast cell proliferation

Nephritis: kidney inflammation, accompanied by proteinuria, haematuria, oedema and hypertension

Nephrotoxic: a substance that is harmful to the kidney

Neuralgia: severe pain along nerve ends

Neuritis: inflammation of nerves

Neuron: nerve cell

Neuropathy: general term for diseases of the central or peripheral nervous system

Neurotoxin: substance with adverse effects on the central and peripheral nervous system; such as transient modulation of mood or performance of CNS

Neurotransmitters: signal compounds in synapses of neurons that help to convert an electric signal into a chemical response; important neurotransmitters are acetylcholine, noradrenaline, adrenaline, dopamine, serotonin, histamine, glycine, GABA, glutamate, endorphins and other peptides

Neurovesicle: small vesicles in the presynapse that are filled with neurotransmitters

Non-protein amino acid (NPAA): secondary metabolite that is an analogue of a proteinogenic amino acid; if incorporated into proteins, the latter are usually inactivated

No-observed-adverse-effect-level (NOAEL): greatest concentration or amount of a substance, which causes no detectable adverse alteration

Noxious substance: harmful substance

Nutritional supplement: a preparation that supplies additional nutrients or active compounds to the body that may not be obtained by the normal diet

Nutritive: nourishing, nutritious

Nycturia: nightly urge to urinate

Nystagmus: involuntary, rapid, rhythmic movement of the eyeball

Oedema (edema): swelling of tissue due to an accumulation of fluids, often caused by kidney or heart failure

Oestrogen (estrogen): a female sex hormone

Ointment: semisolid medicinal preparation that is used topically

Oleum: non-volatile oil; fat

Oliguria: elimination of a small amount of urine in relation to fluid intake

Ophthalmic: relating to the eye

Opium: dried latex of *Papaver somniferum* with several alkaloids, especially morphine

Oral: by mouth (p.o., *per os*)

Organelle: nanomachine or membrane-embraced compartment within a cell that has a specialised function, e.g., ribosome, peroxisome, lysosome, Golgi apparatus, mitochondrion and nucleus

Organic (bio-organic): terms used for products that are grown and processed without the use of artificial chemicals (wild-harvested materials usually qualify as organic)

Organic compound: molecules made of carbon and hydrogen atoms, often containing oxygen and nitrogen in addition

Osteoporosis: a reduction in bone mass, resulting in fractures

OTC: over the counter, a drug that is sold without prescription

Otitis: inflammation of the ear

Oxidation: a reaction that adds oxygen atoms to a molecule or when electrons are lost to a molecule

Oxytocic: speeding up of parturition

Oxytocin: a hormone of the pituitary gland that stimulates lactation and induces labour

Pancytopenia: deficiency of all three cellular components of the blood (red cells, white cells, and platelets).

Panicle: compound flower cluster comprising two or more racemes on a common stalk

Palpitation: noticeable regular or irregular heartbeat

Paraesthesia: abnormal sensation, as burning or prickling

Paralysis: loss or impairment of muscle activity

Parasympathetic nervous system: that part of the nervous system that slows the heart rate, increases intestinal (smooth muscle) and gland activity, and relaxes sphincter muscles

Parasympatholytic: anticholinergic substance that induces effects resembling those caused by interruption of the parasympathetic nerve (e.g. tropane alkaloids)

Parasympathomimetic: cholinomimetic substance that induces effects resembling those caused by stimulation of the parasympathetic nervous system

Parenteral administration: administration of medicinal substances by injection (i.v. = intravenous; i.m. = intramuscular; s.c. = subcutaneous) or intravenous drip

Paresis: weak paralysis

Parkinson's disease: a progressive neurological disease (caused by a degeneration of the *Substantia nigra* and a reduction of dopamine concentrations) marked by lack of muscular coordination and mental deterioration

Parkinsonism: one of several neurological disorders manifesting in unnaturally slow or rigid movements

Pathogen: a microorganism that may cause disease

Percutaneous: absorption through the skin

Periodontitis: inflammation of the area around a tooth

Peristalsis: waves of involuntary contraction in the digestive system

Peritoneal dialysis: clinical procedure of artificial detoxification in which a toxic substance from the body is absorbed into a liquid that has been pumped into the peritoneum

Pesticide: substance (such as fungicide, insecticide, herbicide) used to kill agricultural pests and pathogens

Petal: each of the segments of the corolla of a flower, which are modified leaves and are typically brightly coloured

***p*-gp:** *p*-glycoprotein; an important ATP-driven transporter at biomembranes, which pumps out lipophilic xenobiotics

Pharmaceuticals: drugs, medical products, medicines, or medicaments

Pharmacodynamics: the study of how active substances work in the body; e.g., whether they bind to a receptor

Pharmacogenetics: the study of the influence of hereditary genetic factors on the effects of drugs on individual organisms

Pharmacognosy: the study of herbal drugs, their identification, properties and uses

Pharmacokinetics: the study of how active substances are absorbed, moved, distributed, metabolised and excreted

Pharmacology: the study of the nature, properties and uses of drugs (see pharmacodynamics, pharmacokinetics); includes the study of endogenous active compounds

Pharmacopoeia (pharmacopeia): an official, authoritative publication listing all the various drugs that may be used

Phase 1 reaction: enzymic modification of a xenobiotic or drug by oxidation, reduction, hydrolysis, hydration, dehydrochlorination or other reactions

Phase 2 reaction: conjugation of a substance, or its metabolites from a phase 1 reaction, with endogenous hydrophilic molecules, making them more water-soluble that may be excreted in the urine or bile

Phenotype: the expressed structural and functional characteristics which depend on genotype and environmental conditions

Phlegm: catarrhal secretion or sputum

Phorbolester: diterpene from Euphorbiaceae and Thymelaeaceae, resembling diacylglycerol in structure and therefore activates protein kinase C

Phosphodiesterase: enzyme of signal transduction; inactivates cAMP or cGMP

Phospholipase C: enzyme of signal transduction; splits inositol phosphates to IP3 and diacylglycerol (DAG)

Phospholipids: phosphorylated lipids that are building blocks of cell membranes

Photochemotherapy: the use of phototoxic (UV-activated) furanocoumarins to treat skin ailments

Photo-irritation: skin inflammation due to light exposure, caused by metabolites which were formed in the skin by photolysis

Photosensitisation: increasing sensitivity to sunlight

Phototoxicity: adverse effects of compounds activated by light exposure (e.g. furanocoumarins)

Phytomedicine: *see* Phytotherapy

Phytotherapy: application of plant drugs or products derived from them to cure diseases or to relieve their symptoms

Phytotoxicity: adverse effects of compounds to plants

Pinnate: compound leaf with a main axis and one or more pairs of leaflets plus a terminal leaflet

Piscicide: toxin used to kill fish

Placebo: drug preparation without active ingredients, which cannot be distinguished from the original drug; used in placebo-controlled clinical trials

Placebo effect: an improvement of a health condition that cannot be ascribed to the treatment used

Plasma (blood plasma): cell-free liquid part of blood which surrounds the blood cells and platelets

PMS: premenstrual syndrome; can occur a few days before menstruation with symptoms of irritability, changing mood, insomnia, headache, swollen breasts, cramping and oedema

Pneumonitis: inflammation of the lung

Point mutation: exchange of a single base pair in DNA

Poison: toxicant that causes immediate death or illness even at very low doses

Potentiation: the response to a combination of active substances is greater than expected from the individual compounds (synergism)

Poultice: a semisolid mass of plant materials in oil or water applied to the skin

ppb: parts per billion, 1 µg in 1 kg

ppm: parts per million, 1 mg in 1 kg

Precursor: substance from which another molecule is formed

Prescription drug: a drug that requires a prescription from a physician

Procarcinogen: compound which has to be metabolised before it can induce a tumour

Prodrug: a substance that is converted to its active form within the body

Prophylactic: a substance that prevents disease

Prostaglandins: a group of physiologically active substances within tissues that cause stimulation of muscles and numerous other metabolic effects; important for inflammation processes

Prostate: a gland at the base of the male bladder that secretes a fluid that forms part of semen (stimulating sperm motility)

Prostatitis: bacterial infection of the prostate (*also see* Benign prostate hyperplasia)

Protein kinases: enzymes that phosphorylate other proteins which become activated or inactivated by this modification; important are protein kinase A and protein kinase C

Proteinuria: excretion of excessive amounts of protein in the urine

Pruritus: itching

Psoriasis: inherited skin condition caused by an enhanced growth of dermal cells (keratocytes) resulting in the production of dandruff

Psychosis: mental disorder characterised by personality changes and loss of contact with reality

Psychotropic: a substance that affects the mind or mood

Pulmonary: referring to the lungs

Purgative: *see* Laxative

PUVA: therapy for eczema, psoriasis and other skin ailments combining the drug psoralen and long-wave ultraviolet light

Pyretic (pyrogen): a substance that induces fever

Raceme: flower cluster of stalked flowers, with the youngest ones at the tip

Radix: dried roots

Raphe: a longitudinal ridge on the side of some seeds

Rate: frequency of events during a specified time interval

Reactive oxygen species, ROS: *see* Antioxidant; ROS are thought to be involved in the development of several health disorders (arteriosclerosis, cancer, dementia) and ageing

Receptor: protein (often a membrane protein) that has a binding site for another molecule ("ligand"); important for signal transduction in cells

Relaxant: a substance that reduces tension

Renal: referring to the kidneys

Repellent: compounds used to repel herbivores or predators

Replication: duplication of DNA prior to cell division

Reproductive toxicology: adverse effects of chemicals on the embryo, foetus, neonate and prepubertal mammal and the adult reproductive and neuro-endocrine systems

Resin: amorphous brittle substance resulting from a plant secretion

Resina: resin

Resorption: uptake of a substance through the skin or a mucous membrane (*see* Absorption)

Re-uptake inhibitor: inhibitors of transporters for the neurotransmitters dopamine, noradrenaline and serotonin at presynaptic and vesicle membrane

Rheumatism: general term referring to painful joints

Rhinitis: inflammation of the mucosa of the nose

Rhizoma: dried rhizomes; underground stem

Ringworm: a fungal infection of skin

Risk: probability of manifestation of a hazard under specific conditions

Roborant: tonic or strengthening mixture

Rodenticide: pesticide used to kill rodents

Rubefacient: a substance (counter-irritant) that causes reddening of skin

Saluretic: a substance that increases the concentration of salts in urine

Index

Plant names and page numbers in **bold** indicate main entries; page numbers in ***bold italics*** indicate illustrated plants or chemical compounds.

abedul 120
Abendländischer Lebensbaum 259
abhaya 258
Abies 56
abietic acid ***56***
abrin 27, 91, ***93***
Abrus precatorius 27, ***91***, **93**
abrusosides 93
Absinth 113
absinthe 113
absinthe chinoise 114
Absinthii herba 113
absinthin ***113***
Abyssinischer Tee 130
acacia del Senegal 93
Acacia senegal 9, 29, **93**
Acacia spp. 75
acacie gomme arabique 93
acai berry 18, **163**
acai oil 76
Açaí-Beere 163
açaizeiro 163
Acanthaceae 85, 86, 106, 188
acemannan **75**, **102**
acetaldehyde 109
acetic acid ***76***
acetlacteol 139
acetylandromedol 57
acetylcholine 264
acetyldigoxin 20
acetylindicine 177
2′-acetylneriifoline 133
achillea millefoglio 94
Achillea millefolium 94
Achillea spp. 54
achillicin 94
Achyranthes spp. 58
ackee 89, 121
Ackerschachtelhalm 160
Acmella oleracea 251
Acokanthera oppositifolia 25, ***27***
aconit napel 94
aconite 10, **94**
aconitine 20, 23, 26, ***79***, ***94***, 177
aconito 94
Aconitum napellus 10, 20, 23, 26, **94**
Aconitum spp. 79
Acoraceae 95
acore vrai 95
acorenone **95**
Acorus calamus 95
acridone alkaloids 86
acronine ***86***
Acronychia baueri 86
acronycine 86
Actaea racemosa 16, **139**
actaein ***139***
actée à grappet 139
acylglycerides 29
Adansonia digitata 19, **95**
adansonie d' Afrique 95
adelfa 206
Adenia digitata 27

Adenium multiflorum 96
Adenium obesum 9, 26, **96**
Adenium obesum var. *multiflorum* 96
Adenium obesum var. *obesum* 96
Adhatoda vasica 188
Adhatodae vasicae folium 188
Adlerfarn 230
adonide 96
adonide du printemps 96
Adonidis herba 96
Adonis spp. 62
Adonis vernalis 20, **96**
adoniside 20
Adoxaceae 241
adulterants 40
aerial parts 28
aescin 20, 97
Aesculus hippocastanum 17, 20, 66, **97**
aetheroleum 29
Aethusa cynapium 138
Affenbrotbaum 95
aflatoxin 25
Aframomum melegueta 97
African cherry 9
African Herbal Pharmacopoeia 9
African horehound 119
African medicine 8
African myrrh 143
African potato 9, 18
African rue 22, **214**
African Traditional Medicine 8, 32
African wild olive 209
African wormwood 9, **113**
Afrikanische Teufelskralle 175
Afrikanischer Wermut 113
Agathosma betulina 9, **98**
Agathosma crenulata 98
agave 71
Agave tequilana 71
aglio 101
aglio orsino 101
Agni casti fructus 270
agnocasto 270
agracejo 120
agrimonia 98
Agrimonia eupatoria 20, 70, **98**
Agrimonia pilosa 98
Agrimonia procera 98
Agrimoniae herba 98
agrimoniin **70**, 98, 99, ***166***, 226, 242
agrimony 70
agrimophol 20
agripalma 192
agripaume 192
agroclavine ***80***
Agropyri repentis rhizoma 159
Agropyron repens 159
Agrostemma githago 99

agrostin 99
aigremoine gariot 98
ail blanc 101
ail des ours 101
airelle à gros fruits 264
airelle myrtille 265
airelle rouge 265
Aizoaceae 244
ajacine 26
ajenjo 113
ajmalicine 20, ***81***
ajmaline 20, ***81***
ajo 101
ajo de oso 101
ajowan 12
Ajuga spp. 58
ajugol ***192***
ajugoside 192
akee 121
akee d'Afrique 121
Akeepflaume 121
alanine ***88***
alantolactone ***54***, ***185***
alcachofera 150
alcanfor 140
alcaravea 129
Alcea rosea 104
Alcea spp. 74
alchemilla 99
Alchemilla spp. 70
Alchemilla vulgaris 99
Alchemilla xanthochlora 99
Alchemillae herba 99
alchimille 99
alcohol 22, 23, 271
alder buckthorn 233
alearrhofa 150
Alepidea 57
Aleppo rue ***48***, 239
Aleurites fordii 100
alevrite 100
alfa-alfa 200
alfalfa 200
algarrobo 133
algin 115
alginate 115
alginic acid 115
alkaloids 16, 36, 44, 78–87
alkamides 16, 17, 73
allantoin 20, ***254***
Alliaceae 89, 100, 101
allicin ***89***, ***101***
Allii cepae bulbus 100
Allii sativi bulbus 28, 101
Allii sativi pulvus 101
Allii ursini bulbus 101
Allii ursini herba 101
alliin **89**, 101
Allium cepa 12, 17, 28, 89, **100**
Allium sativum 16, 28, **101**
Allium ursinum 19, **101**, 144
allyl isothiocyanate 20, ***90***, ***111***
almendro 227
almond 12, **227**
almorta 190

Aloe barbadensis 102
Aloe capensis 102
aloe del Capo 102
Aloe ferox 9, 19, 29, **102**
Aloe spp. 72
Aloe vera 12, ***13***, 18, 29, **102**
aloe vera 18, **102**
Aloe vera gel 75, 123
aloe-emodin 234
aloe-emodin anthrone 102
aloes feroce 102
aloes vrai 102
aloin ***72***, ***73***, ***102***
Aloysia citrodora 19, 51, **103**
Aloysia triphylla 103
Alpinia galanga 103
Alpinia officinarum 103
Alraune 198
Alstroemeria spp. 71
altea 104
Althaea officinalis 75, **104**
Althaeae radix 104
amabiline 122
amalika 218
amamelide 174
amandier 227
Amanita muscaria 22, ***25***, 26
Amanita phalloides 26
amanitin 26
amaranth seed oil 58
Amaranthaceae 60, 135
amarogentin ***52***, *170*
amarum 102, 131, 132, 202, 231
Amaryllidaceae 78, 100, 101, 121, 168, 205
amaryllidaceae alkaloids 78, 168
ambelline 78
Ambrosia elatior 91
ameo mayor 104
American arbor-vitae 259
American aspen 225
American blueberry 265
American cranberry 264
American ginseng 212
American mandrake 223
American spikenard 15, 18, **108**
Amerikanische Espe 225
Amerikanische Narde 108
Amerikanische Traubensilberkerze 139
Amerikanischer Faulbaum 234
amino acids 88
α-amino-β-oxalyl aminopropionic acid 89
amiomaior 104
amio-vulgar 104
amla 218
amlak 218
Ammi majus 21, **104**
Ammi visnaga 270
Ammi visnagae fructus 270
ammodendrine ***84***
3,11-amorphadien 164
Amorphophallus konjac 75

ampelopsin ***180***
amphetamine ***84***, 130
amritoside 229
Amsinckia 85
Amur cork tree 217
amurensin 217
Amur-Korkbaum 217
amygdalin ***90***, ***227***, ***228***
anabasine 20, 156, ***207***
Anabasis aphylla 20
Anacardiaceae 65, 235, 245
Anadenanthera peregrina 14, 23, 78, **105**
anagyrine ***85***
analysis of compounds 36
Anamirta cocculus 21, 54
Ananas 105
Ananas comosus 19, 20, 91, **105**
Ananas sativa 105
ananasso 105
androgens 58
Andrographidis paniculatae herba 106
Andrographis paniculata 13, 20, 21, **106**
andrographolide 20, ***106***
andromedotoxin ***57***
Anemone spp. 71
aneth 106
Anethi aetheroleum 106
Anethi fructus 106
Anethi herba 106
anethole 50, ***64***, 65
trans-anethole ***166***, ***184***, ***219***
Anethum graveolens 106
Anethum sowa 106
aneto 106
angel's trumpet 14, 22, 25, **123**
angelic acid ***77***
angelica 107
angelica 66, 77, 107
Angelica archangelica 66, 77, **107**
Angelica atropurpurea 107
Angelica dahurica 107
Angelica sinensis 12, **107**
Angelicae fructus 107
Angelicae herba 107
Angelicae radix 107
Angelicae sinensis radix 107
angelicin ***66***, ***107***
angelique Chinoise 107
angelique de Chine 107
anice stellato 184
anice verde 219
Anis 219
anis étoilé 184
anis vert 219
p-anisaldehyde ***76***
anisatin ***54***, 184
anise 29, **219**
Anisi fructus 29
Anisi stellati frutus 184
anisillo 255
anisodamine 20

289

anisodine 20
Anisodus tanguticus 20
Ankyropetalum gypsophiloides 60
Annonaceae 82
Anserinae herba 225
Anthemis nobilis 134
anthocyanins 17, 18, 44, 67–69, 178, 163, 264, 265, 271
Anthoxanthum odoratum 66
anthraquinones 72, 233
Anthriscus 67
anthroposophical medicine 10
antiarin 27
Antiaris toxicaria 27
aphrodisiac 10
Apiaceae 28, 50, 52, 54, 57, 66, 67, 73, 77, 104, 106, 107, 108, 129, 132, 138, 144, 145, 152, 165, 166, 193, 216, 219, 220, 243, 270
apigenin *67*, 213
Apii fructus 108
apiin 216
apio 108
apiol *64*, 108, ***216***
Apium graveolens **108**
Apocynaceae 52, 62, 63, 81, 82, 86, 96, 133, 179, 203, 206, 232, 252, 259, 268, 272
Apocynum spp. 62
apomorphine 20
aporphine 82
apples 71
apricot vine 213
Aquifoliaceae 183
Arabian coffee 142
arabino-galactans, 260
arabinose *93*, 104
Araceae 76, 114, 154
Arachis hypogaea 91
arak 240
Aralia elata 108
Aralia japonica 60
Aralia mandshurica 108
Aralia racemosa 15, 18, **108**
Araliaceae 60, 73, 108, 159, 176, 212
araloside A ***60***, ***108***
arancio amaro 141
arandano commun 265
arbol de las perlas 204
arbol de los escudos 170
arbol de seso 121
arbre à lait 259
arbre aux serpents 232
arbre fricasse 121
arbre-a-encens 122
arbutin *72*, **109**, 169, 262
archangel 107
archangelica 107
archangelique 107
arctic root 235
Arctii radix 109
arctinal ***73***, **109**
Arctium lappa 73, **109**
Arctium majus 109
Arctium minus 109
Arctium tomentosum 109
Arctopus spp. 57
Arctostaphylos uva-ursi 109

Ardisia japonica 20
Areca catechu 13, 20, 22, 84, **110**
areca nut 110
Arecaceae 110, 163, 248
arecaidine ***84***, 110
arecoline 20, 22, ***84***, **110**
arenarin 176
arequier 110
Argemone mexicana 110
Argemone ochroleuca 23, **110**
argentina anserina 225
argentine 225
arginine ***88***, ***217***
argousier 179
Argyreia 80
arhar 124
aristolochia 111
Aristolochia clematitis **111**
Aristolochiaceae 73, 111
Aristolochiae herba 111
aristolochic acid I ***111***
aristoloquia 111
Aristotle 10
armoise d'Afrique 113
Armoracia lapathifolia 111
Armoracia rusticana 90, **111**
Armoraciae radix 111
arnica 10, **112**
Arnica chamissonis 54, 112
Arnica fulgens 112
Arnica montana 10, 54, **112**
Arnicae flos 112
Arnika 112
aro gigaro 114
aroeira 245
aroin 26
aromatherapy 10, 29, 32, 50
aromatic acids 76
aronga 175
Aronia 69
Aronia melanocarpa 19, **112**
aronie a fruits noirs 112
Aronstab 114
arpagofito 175
arreteboeuf 210
artabsin 113
artemether 20, 114
Artemisia absinthium 23, 51, 53, **113**
Artemisia afra 9, **113**
Artemisia annua 12, 16, 20, 54, **114**
Artemisia argyi 20
Artemisia cina 55
Artemisia maritima 21, 55
Artemisia sp. 35
Artemisia vulgaris ***55***, 113
Artemisiae africanae herba 113
artemisinin 16, 20, 42, ***54***, ***114***
artesunate 54
artichaut 150
artiglio del diavolo 175
Artischocke 150
Arum maculatum 26, **114**
arum tachete 114
arveira 245
Arzneibaldrian 266
asafoetida 12, ***13***, **165**
Asant 165

β-asarone 95
Asarum europaeum 111
ascaridol *50*, 51, ***135***
Asclepiadaceae 179, 272
***Asclepias tuberosa* 59**
ascophylle 115
Ascophyllum nodosum 19, **115**
ascorbic acid 74, ***77***, 95
ase fétide 165
ashwagandha 16, 22, **272**
Asian ginseng 212
asiatic acid ***132***
Asiatic flea seeds 221
asiaticoside 17, 20, ***60***, 132
Asiatischer Wassernabel 132
Aspalathi linearis herba 115
aspalathin ***68***, ***115***
aspalathus 115
Aspalathus linearis 9, 19, 33, 68, **115**
Asparagaceae 116, 121, 144, 155, 238
Asparagi radix 116
asparagine ***88***, ***116***, 217
asparago 116
asparagus 116
Asparagus officinalis **116**
aspartic acid ***88***
asperge 116
aspergillus 25
Aspergillus flavus 25
asperuloside 222
Asphodelaceae 72, 102, 123
aspirin 10, 20, 42, 165, 239
assafetida 165
assenzio 113
Asteraceae 50, 52, 54, 57, 63, 66, 67, 73, 74, 85, 94, 109, 112, 113, 114, 124, 125, 129, 131, 134, 137, 138, 150, 157, 158, 173, 176, 185, 199, 247, 249, 251, 252, 255, 256, 257, 263
asthma weed 194
Astracantha gummifera 12
astragale 116
Astragali radix 116
astragalo 116
astragaloside A ***116***
astragalus 116
Astragalus gummiferum 75
Astragalus membranaceus **116**
Astragalus spp. 82
Atractylis gummifera 57
atractyloside ***57***, ***125***
Atropa belladonna 10, 20, 22, 26, ***48***, **117**
Atropa spp. 87
atropine 10, 20, 22, 23, 26, 27, 35, ***87***, 117, 151, 182, 246
atroscine 246
aubépine 146
aubour 189
aucubin ***52***, ***221***, 222, 267
Aufgeblasene Lobelie 194
aunée 185
Aurantii pericarpum 29, 141
aurones 67
Australian chestnut ***79***, 82

Australian lemon myrtle 51
Australian traditional medicine 14
autumn crocus 25, 28, ***79***, **143**
avellana d'India 110
avena 117
Avena sativa 19, ***117***
Avenae 80
Avenae fructus 117
Avenae herba recens 117
Avenae stramentum 117
Avenzoar 12
Avicenna 12
avorniello 189
avoine 117
ayahuasca 14, 22, **119**
Ayurvedic medicine 12–13, 47, 86
Azadirachta indica 13, **118**
azadirachtin ***118***
azafrán 143
azulejo 131
azulenes 134
ba jiao xian 184
baby's breath 174
bacc 95
Bach flower remedies 10, 11
Bach, Edward 11
Backhousia citriodora 51
Bacopa 118
bacopa de Monnier 118
Bacopa monnierii **118**
bacopaside 118
bacoside A & B ***118***
badiane de Chine 184
badiane du Japon 184
Badiano, Juan 14
bai shao yao 12, 211
bai zhi 107
baicalin ***246***
Baikal skullcap 246
Baikal-Helmkraut 246
Ballonblume 222
balloon flower 222
Ballota africana 119
Ballota foetida 119
ballota nera 119
Ballota nigra 56, **119**
Ballotae nigrae herba 119
ballote 119
ballotenol ***56***, ***119***
balm of Gilead 29, 225
balota 119
balsam poplar 225
Balsam tolutanum 205
Balsaminaceae 72
balsamo del Tolu 205
balsams 29
Banisteriopsis argentea 78
Banisteriopsis caapi 14, 22, **119**
Banisteriopsis rusbyana 78
Banisteriopsis spp. 81
baobab 19, **95**
baobab africano 95
baobab del Africa 95
Barbados aloes 102
barbaloin 102
barbarasco 267

bardana 109
Bardanae folium 109
Bardanae radix 109
bark 28
Barlauch 101
barley 16, 19, **180**
Barosma betulina 98
Barosmae folium 98
Basilic sacre 208
basket willow 239
bastard star anise 184
bathing 34
baumier de Tolu 205
Baumtabak 207
bear's garlic 19, **101**
bearberry 109
Beech, Wooster 14
been 204
Behaartenkraut 136
behen oil 19, 204
behenic acid ***204***
belladonna 117
belladonna 117
Belladonnae folium 117
Belladonnae radix 117
belladonne 117
ben 204
Ben Cao Gang Mu 12
ben oil 19, 204
ben oleifere 204
ben tree 204
Benediktenkraut 131
bengkudu 203
benzaldehyde 109
benzodiazepines 221
benzoic acid 29, 76, 77, 124, 205
benzoin 14, ***15***
benzyl benzoate 20, ***205***
benzyl glucosinolate 262
benzyl isothiocyanate ***240***, 262
Berberidaceae 67, 82, 120, 223
Berberidis cortex 120
Berberidis fructus 120
Berberidis radici cortex 120
Berberidis radix 120
berberine 20, ***82***, 110, ***120***, 135, ***145***, 181, 197, ***217***
Berberis aquifolium 197
berberis commun 120
Berberis vulgaris 20, **120**
berbero 120
bergamot oil 141
bergapten ***66***, ***104***, 153
bergenin 20
Bergwohlverleih 112
berro de huerta 192
Besenginster 150
beta glucan 18
Beta vulgaris 71
betacyanins 68
betalains 68
betanidin 69
betanin ***69***
betel nut 13, 22, 35, 84, 110
betel vine 22, 110
Betelnusspalme 110
betonicum 94
Betula alba 20
Betula pendula **120**
Betulaceae 120

290